Second Edition

Origin and Evolution of the Universe

From Big Bang to ExoBiology

Second Edition

Origin and Evolution of the Universe

From Big Bang to ExoBiology

editors

Matthew A Malkan
University of California, Los Angeles, USA

Ben Zuckerman
University of California, Los Angeles, USA

NEW JERSEY • LONDON • SINGAPORE • BEIJING • SHANGHAI • HONG KONG • TAIPEI • CHENNAI • TOKYO

Published by

World Scientific Publishing Co. Pte. Ltd.
5 Toh Tuck Link, Singapore 596224
USA office: 27 Warren Street, Suite 401-402, Hackensack, NJ 07601
UK office: 57 Shelton Street, Covent Garden, London WC2H 9HE

British Library Cataloguing-in-Publication Data
A catalogue record for this book is available from the British Library.

ORIGIN AND EVOLUTION OF THE UNIVERSE
From Big Bang to ExoBiology
Second Edition

Copyright © 2020 by World Scientific Publishing Co. Pte. Ltd.

All rights reserved. This book, or parts thereof, may not be reproduced in any form or by any means, electronic or mechanical, including photocopying, recording or any information storage and retrieval system now known or to be invented, without written permission from the publisher.

For photocopying of material in this volume, please pay a copying fee through the Copyright Clearance Center, Inc., 222 Rosewood Drive, Danvers, MA 01923, USA. In this case permission to photocopy is not required from the publisher.

ISBN 978-981-120-645-0 (hardcover)
ISBN 978-981-120-772-3 (paperback)
ISBN 978-981-120-646-7 (ebook for institutions)
ISBN 978-981-120-647-4 (ebook for individuals)

For any available supplementary material, please visit
https://www.worldscientific.com/worldscibooks/10.1142/11447#t=suppl

Desk Editor: Rhaimie Wahap

Typeset by Stallion Press
Email: enquiries@stallionpress.com

Contents

Preface	vii
Editors	ix
Authors	xi
Chapter 1 The Origin of the Universe *Edward L. Wright*	1
Chapter 2 The Origin and Evolution of Galaxies *Alan Dressler*	25
Chapter 3 The Origin and Evolution of the Chemical Elements *Virginia Trimble*	63
Chapter 4 Stellar Explosions, Neutron Stars, and Black Holes *Alexei V. Filippenko*	99
Chapter 5 The Origin of Stars and Planets *Fred C. Adams*	149
Chapter 6 The Origin and Evolution of Life in the Universe *Christopher P. McKay*	189
Glossary	213
Index	227

Preface

In March 1995, we had the privilege and pleasure of organizing a Symposium at UCLA on "The Origin and Evolution of the Universe". It attracted overflow audiences from diverse elements of the Los Angeles community, including college students, faculty, researchers and members of the interested general public. The speakers then carefully prepared and extended their presentations for our 1996 publication of the First Edition of "The Origin and Evolution of the Universe".

The subject of the original book is still as fresh and exciting as it ever has been. Now there are several compelling reasons we decided to produce a thoroughly revised second edition.

Discoveries and advances have unfolded at such a remarkable pace, exceeding even our most optimistic expectations. In 1996, supernovae had not yet been used to establish a cosmic distance scale accurate enough to determine that the Universal expansion is — surprisingly — accelerating. The remarkable detection of gravitational waves from merging black holes and neutron stars by LIGO was still decades in the future. And essentially no planets beyond our solar system were known to exist, and certainly none that were anything like our Earth. Today everything has changed.

In 1996, the authors of each chapter were already prominent researchers in their fields. Over the last two decades, their careers continued to advance and even accelerate. Today they are each an undisputed active leader in their area of scientific specialization. They are also in demand as skilled communicators of science to the wider public. They are thus ideal writers to convey the latest major developments in astronomy.

In addition, since 1996, the importance of these big themes of science to non-scientists has arguably grown even larger. Scientific research is now depending on public support for even larger amounts of funding than ever before. Many specific successful examples appear throughout the chapters in this book. Scientific results from the European Space Agency's Planck and Newton/XMM satellites are described in Chapters 1 and 4. New exoplanet discoveries from NASA's Kepler satellite are described in Chapter 6. And further scientific breakthroughs from the multi-national collaborations to build and operate the Hubble Space Telescope, as well as ALMA — a giant array of radio telesopes located in northern Chile — are described in Chapters 2 and 5. We are already anticipating comparable publicly-funded advances in astronomy with the James Webb Space Telescope, scheduled for launch in 2021.

One broader point should be clear to readers of this Second Edition. In the entire Universe, Earth's biosphere remains the only home of life that we know of. Whether or not we are prepared for it, we humans have thus inherited a tremendous responsibility as Earth's stewards.

As in 1996, many people today regard the climax of the evolution of the Universe to be life, and especially "intelligent" life. It remains an open question whether or not this will be enough. "Intelligence" makes our technology possible, but it should be much more. Are we truly "intelligent", or only "technological"? Human intelligence notwithstanding, we often seem to be our own worst enemy. Our rapid technical advances have often not been matched by comparable improvements in our ability to get along with each other and our environment. Using our *full* human intelligence is our best hope for the future. Our species will have to be smarter — and *act* smarter. Our success or failure at solving our problems in managing our uniquely precious planet provide the sharpest test yet of real human intelligence. If we can pass this test, then we may also be able to answer the outstanding future questions explored and raised in this book.

Matthew Malkan and Benjamin Zuckerman
UCLA

Editors

Matthew Malkan
Matthew Malkan is a Distinguished Professor of Physics and Astronomy at UCLA. After graduating Harvard with summa cum laude and Phi Beta Kappa honors, Dr. Matthew Malkan studied at the University of Cambridge as a Marshall Scholar. He received his PhD from Caltech on a Hertz Fellowship, and was a postdoctoral research fellow at the University of Arizona. Upon starting his Assistant Professorship at UCLA, Dr. Malkan was a Presidential Young Investigator from 1986 to 1991. He has been Professor of Physics and Astronomy at UCLA since 1992. In 2009, Malkan was the AMC-FUMEC Distinguished Visiting Professor in Mexico. He has published over 450 refereed articles in peer-reviewed journals, and has worked extensively on astronomy-related film and television shows, behind and in front of the camera. Malkan uses a wide range of telescopes in space and on the ground to study the evolution of galaxies and their massive black holes.

Ben Zuckerman
Ben Zuckerman is a Professor in the Dept. of Physics & Astronomy at UCLA. He received undergraduate and graduate degrees from MIT and Harvard. His major scientific interests have been the birth and death of stars and planetary systems. He has maintained a continuing interest in the question of the prevalence of life — especially intelligent life — in the Universe and, beginning

way back in the mid-1970s, developed and regularly taught a course on "Life in the Universe". He also developed and taught a UCLA Honors course entitled "The 21st Century: Society, Environment, Ethics". He believes that while our species (Homo sapiens) may be considered to be "technological", because we are destroying our home — Earth's biosphere — we cannot be considered to be "intelligent". Zuckerman has co-edited six books including, *Extraterrestrials, Where Are They?* (Cambridge University Press) and *Human Population and the Environmental Crisis* (Jones & Bartlett).

Authors

Edward L. Wright
Edward L. Wright has been a professor at UCLA since 1981. Wright works in infrared astronomy and cosmology. As an interdisciplinary scientist on the Space InfraRed Telescope Facility (SIRTF) Science Working Group, Wright has worked on the SIRTF project (renamed the Spitzer Space Telescope) since 1976. He was an active member of the teams working on the Cosmic Background Explorer (COBE) since 1978. Wright is the principal investigator of the Wide-field Infrared Survey Explorer (WISE) project. Wright is also a member of the science team on the Wilkinson Microwave Anisotropy Probe (WMAP), which launched in June 2001. Prof. Wright was elected to the US National Academy of Sciences in 2011. Wright was named the Distinguished Scientist of the Year, in 1995 by the Center for the Study of the Evolution and Origin of Life. In 1992 he received the NASA Exceptional Scientific Achievement Medal for his work on COBE, and NASA's highest honor, the Distinguished Public Service Medal, in 2018. He received the Breakthrough Prize in Fundamental Physics as part of the WMAP team in December 2017.

Alan Dressler

Dressler is Astronomer Emeritus at the Observatories of the Carnegie Institution for Science, in Pasadena, California. Dressler's principal area of research is the formation and evolution of galaxies. He makes observations with large ground- and space-based telescopes — imaging and spectroscopy, to study the morphological types, structures, stars, and kinematics of galaxies. Dressler investigates galaxy evolution as it happened — by observing galaxies so distant that they are seen as they appeared billions of years ago. A principal direction of his research has been the charting of the diverse histories of star formation in galaxies, a strong constraint on theoretical models of how such structures have developed over cosmic time. Dressler has been an active member of the astronomical community and has served on and chaired many committees, for NSF, NASA, and the National Academy of Sciences, that make recommendations on future facilities like the James Webb Space Telescope and the next generation extremely large telescopes. He is an experienced and passionate promoter and popularizer of science and has written many magazine articles and a book, "Voyage to the Great Attractor: Exploring Extragalactic Space" that puts our studies of the universe in the broader context of humanity's place in the natural world.

Virginia Trimble

Virginia Trimble is a native Californian and graduate of Hollywood High School, UCLA, and Caltech (PhD 1968) with honorary degrees from the Universities of Cambridge UK (MA 1969) and Valencia Spain (Dott. h.c. 2010). She is the oldest faculty member in physics and astronomy of the University of California, Irvine still on full active duty (having been the youngest in 1971, the only woman and the only astronomer for the first 15 years). Trimble is the only person to have been President of two different Divisions of the International Astronomical Union (Galaxies and the Universe; Union-Wide Activities) and is currently President of its Commission on Binary and Multiple Stars. Her publication list recently passed number 850, not including the present chapter and a dozen other items under review or in press. Most of her current research is in history of science and scientometrics, after years of plugging away at white dwarfs, supernovae, nucleosynthesis, binary star statistics, and so forth.

Alex Filippenko

Alex Filippenko is a Professor (and currently the Chair) of Astronomy at the University of California, Berkeley, where he is also the Richard and Rhoda Goldman Distinguished Professor in the Physical Sciences and a Senior Miller Fellow in the Miller Institute for Basic Research in Science. An elected member of both the US National Academy of Sciences and the American Academy of Arts and Sciences, he is one of the world's most highly cited (>130,000 citations) astronomers and the recipient of numerous prizes for his scientific research. He was the only person to have been a member of both teams that revealed the accelerating expansion of the Universe, an amazing discovery that was honored with the 2007 Gruber Cosmology Prize and the 2015 Breakthrough Prize in Fundamental Physics to all team members, and the 2011 Nobel Prize in Physics to the teams' leaders. Winner of the most prestigious teaching awards at UC Berkeley and voted the "Best Professor" on campus a record 9 times, in 2006 he was named the Carnegie/CASE National Professor of the Year among doctoral institutions, and in 2010 he received the ASP's Richard H. Emmons Award for undergraduate teaching. He has produced 5 astronomy video courses with The Great Courses, coauthored an award-winning astronomy textbook (5 editions), and appears in more than 100 television documentaries. In 2004, he was awarded the Carl Sagan Prize for Science Popularization. He was selected as one of only two recipients of the 2017 Caltech Distinguished Alumni Award. He makes a hobby of observing total solar eclipses throughout the globe, having seen 16 so far, all successfully.

Fred C. Adams

Born in Redwood City, California, Fred Adams received his undergraduate training in Mathematics and Physics from Iowa State University in 1983 and his PhD in Physics from the University of California, Berkeley, in 1988. For his PhD dissertation research, he received the Robert J. Trumpler Award from the Astronomical Society of the Pacific. After serving as a postdoctoral research fellow at the Harvard-Smithsonian Center for Astrophysics, he joined the faculty in the Physics Department at the University of Michigan in 1991. Adams was promoted to Associate Professor in 1996 and to Full Professor in 2001. He is the recipient of the Helen B. Warner Prize from the American Astronomical Society and the National Science Foundation Young Investigator Award. He has also been awarded both the Excellence in Education Award and the Excellence in Research Award from the College of Literature, Arts, and Sciences at the University of Michigan. In 2002, he was given The Faculty Recognition Award from the University of Michigan. In 2007, he was elected to the Michigan Society of Fellows. In 2014, was elected to be a fellow of the American Physical Society and he was named as the Ta-you Wu Collegiate Professor of Physics at the University of Michigan.

Christopher P. McKay

Chris is a senior scientist with the NASA Ames Research Center. His research focuses on life in extreme environments and the search for life on other worlds in our Solar System. He is also actively involved in planning for future Mars missions including human exploration. Chris been involved in research in Mars-like environments on Earth, traveling to ice-covered lakes in Antarctica, permafrost in the Siberian and Canadian Arctic, many deserts including the Atacama, Namib, & Sahara Deserts to study life in these extreme environments. He was a co-investigator on the Huygens probe to Saturn's moon Titan in 2005, the Mars Phoenix lander mission in 2008, and the Mars Science Laboratory mission, in 2012.

Chapter 1

The Origin of the Universe

Edward L. Wright

Introduction

One of the most significant developments of 20th century natural philosophy has been the acquisition, through astronomical observations and theoretical physics, of a substantial understanding of the earliest moments in the history of the Universe. The best current model for the origin of the Universe is known as the **inflationary scenario** in the **Hot Big Bang** model. This chapter describes the observational underpinnings of the Big Bang theory and some theoretical models based on it. The development uses elementary mathematics because this is the simplest way to describe much of physical science.

Expansion of the Universe

The discovery by Hubble (1929) that distant galaxies are moving away from us with a velocity, V, that is proportional to their distance, D, was the first evidence for an evolving Universe. Observations show that this recessional velocity is

$$V = H_0 D \pm V_p \tag{1}$$

where the coefficient H_0 is known as the **Hubble constant** and V_p the peculiar velocity of the galaxy, which is peculiar in the sense that it is different for each galaxy. Typical values of V_p are 500 km/s, and the recessional velocities V range to several times 100,000 km/s for the most distant visible objects.

The **Hubble law** does not define a center for the Universe, although all but a few of the nearest galaxies seem to be receding from our Milky Way galaxy. Observers on a different galaxy, A, would measure velocities and distances relative to themselves, so we replace V by $V-V_A$ and D by $D-D_A$. The Hubble law for A is

$$V-V_A = H_0(D-D_A), \qquad (2)$$

which is exactly the same as the Hubble law seen from the Milky Way (except for slightly different peculiar velocities), provided that $V_A = H_0 D_A$. Thus, every galaxy whose velocity satisfies the Hubble law will also observe the Hubble law. An observer on any galaxy sees all other galaxies receding; thus, the Big Bang model is "omnicentric."

The idea that the Universe looks the same from any position is codified in:

> The Cosmological Principle: The Universe is **homogeneous and isotropic**.

Note that the Hubble law does not define a new physical interaction that leads to an expansion of the Universe. Instead, it is only an empirical statement about the observed motions of galaxies. Each individual object moves on a path dictated by its initial trajectory and the forces that act on it. Since a uniform distribution of matter produces no net force by symmetry, the only forces felt by an object are caused by the structures around it, and these forces can be computed to good accuracy using the equations of Newtonian mechanics and gravity. On small scales, local forces dominate the motions of objects. For example, the orbit of an electron in an atom is determined by electrostatic forces, and the distance between the electron and the nucleus does not increase with time. The orbits of the planets in our solar system are determined by the gravitational force of the Sun, and the distance between a planet and the Sun does not follow the Hubble law. Note that the individual galaxies in Figure 1.1 do not expand. But the motion of galaxies more distant than 30 million light years is well described by the Hubble law, and this observed fact tells us that the density distribution in the early

The Origin of the Universe

Figure 1.1 Schematic evolution of our expanding universe and its contents. In the earlier stage, at the top, a given volume of space has a high density of matter (yellow objects) and background photons. The latter are shown in blue because they have short wavelengths and high energies. In a later stage — middle — this volume has expanded so that its density of matter has dropped. At the same time, the density of photons has dropped, as their average wavelengths have increased (now shown in green). The photons have moved. However, the relative positions of the matter (yellow galaxies) are preserved because they follow a pure uniform Hubble expansion. In a still later stage of the expansion — bottom frame — this region has further expanded. The space density of matter is still lower, and the density of photons has dropped further. The average wavelength of the photons has increased, corresponding to a lower temperature of this background radiation, so that they are now shown in dark red.

Universe was almost uniform and that the initial peculiar velocities were small.

The recessional velocity of an object is easily measured with use of the **Doppler shift**, which causes the length of electromagnetic waves received from a receding object to be larger than the wavelength at which they were emitted by the object. Because the long wavelength end of the visible spectrum is red light, the Doppler shift of a receding object is called its **redshift**. The observed wavelength λ_{obs} is larger than the emitted wavelength λ_{em} and the ratio

$$\lambda_{obs}/\lambda_{em} = 1 + z \tag{3}$$

defines the redshift z used by astronomers. For velocities that are small compared with the speed of light, the approximation $v = cz$ can be used. The most distant known **quasar** has $z = 7.54$. For this object, the ultraviolet (UV) Lyman α line of hydrogen with $\lambda_{em} = 122$ nm is seen at $\lambda_{obs} = 1$ μm [10^{-6}m] in the near infrared. For such large z, corrections to the Doppler shift formula are needed for velocities approaching the speed of light. Figure 1.2 compares the absorption line spectra of a star and three galaxies at progressively larger distances moving up the figure. The characteristic pattern of absorption lines seen in a local star (with no redshift), can be seen to shift further and further to the red wavelengths. This is predicted by Hubble's law, which says that the more distant a galaxy is, the higher its recession velocity from Earth is. Through the Dopper effect, these increasing velocities are observed as increasing shifts of the spectrum to the red — the increase of redshift with distance.

The Hubble law in Equation (1) applies to the relative velocity between any pair of galaxies. For example, the velocity of galaxy A with respect to galaxy B is $V_{AB}(t_0) = H_0 D_{AB}(t_0)$, where $D_{AB}(t_0)$ is the separation now (the time "now" is denoted t_0) between galaxies A and B. If we consider the separation between A and B after a small time interval Δt, it is

$$D_{AB}(t_0 + \Delta t) = D_{AB}(t_0) + V_{AB}\Delta t = D_{AB}(t_0)(1 + H_0 \Delta t). \tag{4}$$

The time interval Δt must be a small fraction of the age of the Universe, and yet the distance light travels in Δt must be larger than structures like clusters of galaxies in which local forces produce large peculiar velocities. Observations of the Universe show that it is smooth enough on

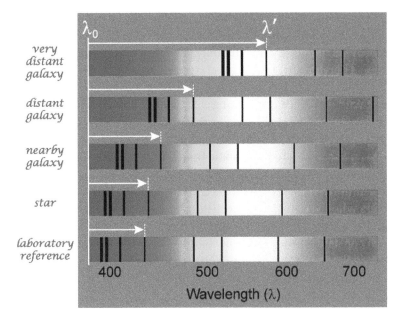

Figure 1.2 Schematic illustration of visible spectra of several objects. The star has a characteristic set of absorption lines (black), which are at almost the wavelengths we see in the laboratory (bottom), because the star has such a small Doppler shift with respect to Earth. The relatively nearby galaxy (middle) shows the same pattern of dark absorption lines as in the star, but all shifted to longer wavelengths. The more distant galaxy, following Hubble's Law, has all of its wavelengths shifted further to the right (to the red). And the most distant galaxy (top) has the largest redshifts. For example the indicated absorption line which is at 440 nanometers in the laboratory (not moving), is shifted all the way to 580 nanometers in the very distant galaxy, and all of its other absorption lines are shifted by this same ratio of 1.32. *Source*: scienceconnected.org.

medium-to-large scales for Equation (4) to be valid. The factor $(1 + H_0 \Delta t)$ is independent of which pair of galaxies A and B is chosen, so it represents a universal scale factor that describes the expansion of every distance between any pair of objects in the Universe. This means that the patterns of galaxies in the Universe retain the same shape while the Universe expands, seen schematically in Figure 1.1. We call the universal scale factor $a(t)$, so

$$a(t_0 + \Delta t) = (1 + H_0 \Delta t) \qquad (5)$$

for times close to the present. Note that $a(t_0) = 1$ by definition.

If there is no acceleration because of gravity, objects will move with constant velocity and Equation (5) is true even if Δt is not small. In this case, when $\Delta t = -1/H_0$, $a(t_0 + \Delta t) = 0$, where a negative Δt denotes an epoch earlier than the present. Thus, all distances in the Universe go to zero at a time $1/H_0$ ago. (By its definition, the Hubble constant has the units of the inverse of time. Therefore it is common to refer to its inverse, $1/H_0$, as the "Hubble Time.") We normally simplify discussions by defining the moment with $a(t) = 0$ (the "Big Bang") to be $t = 0$. This definition makes the age of the Universe equal to the current time, t_0. For the no-acceleration case, $a(t) = t/t_0$ and the product of the Hubble constant and the age of the Universe is $H_0 t_0 = 1$, then the Hubble time is expressed in the same units as the current age of the Universe. In other words, the age of a Universe with no acceleration is always equal to its current Hubble Time. This implies that observers who lived earlier in the history of the Universe, with a smaller t_0, would find a larger Hubble constant H_0. Thus, the Hubble constant is not a physical constant like the electron charge e, because, although the Hubble constant is the same everywhere in the Universe, it changes with time. We call this changing value the Hubble parameter $H(t)$ and define $H_0 = H(t_0)$.

The exact formula for the redshift of an object is $1 + z = a(t_0)/a(t_{em})$, where t_{em} is the time the light was emitted. This states that wavelengths of light expand by exactly the same scale factor that applies to the separations between pairs of galaxies.

The acceleration caused by gravity vanishes only if the Universe is empty, with no mass. When masses are present, gravity provides an attractive force that causes the expansion to slow down. This means that velocities were greater in the past; thus, for a given expansion rate now (H_0), the time since $a = 0$ is smaller than it would have been without any deceleration. In the most likely case, the density of the Universe is very close to the **critical density** that divides underdense Universes that expand forever from overdense universes that will eventually stop expanding and recollapse.

When a small object of mass m is moving under the influence of gravity near a large mass M, the equation that relates its velocity V and distance r from the large mass is

$$E = \tfrac{1}{2} mV^2 - GMm/r, \tag{6}$$

where E is the total energy, which is conserved, $\tfrac{1}{2}mV^2$ is the kinetic energy, and $-GMm/r$ is the gravitational potential energy. Here G is the constant of gravitational force. We can use this simple equation in **cosmology**, with m being a galaxy and M being the mass of the Universe within radius r, which is the density ρ times the volume of a sphere $(4\pi/3)r^3$. The sphere is centered at $r = 0$, and the galaxy m is located on its surface. (*Proving* that we can use this equation requires general relativity.) Because all matter at larger distances than r has larger velocities than $H_0 r$, the matter outside the sphere stays outside. Newton showed that the gravitational force on m from matter outside the sphere is zero, and this is still true under general relativity. Because all matter at smaller distances than r has smaller velocities than $H_0 r$, the matter inside the sphere stays inside. Thus, the mass of the sphere is constant. For a body to just barely escape from r to ∞ requires a total energy $E = 0$. This gives the formula for the **escape velocity**, $v_{esc} = \sqrt{(2GM/r)}$. When the Universe has the critical density, the Hubble velocity $H_0 r$ is equal to the escape velocity, which gives an equation for the mass M leading to the critical density as follows:

$$\rho_{crit} = 3H_0^2/8\pi G. \tag{7}$$

If the Universe has the critical density now, it must have the critical density at all times. Thus, if we can figure out how the density changes as the Universe grows, we can figure out how the Hubble parameter $H(t)$ changes as the Universe grows. For normal matter the density drops by a factor of 8 when the Universe doubles in size. The radiation filling the Universe also contributes to the density, but this density goes down faster than the matter density due to the redshift, dropping by a factor of 16 as the Universe doubles in size. For a critical density Universe, these factors lead to a dimensionless product of the Hubble constant times the age of the Universe $H_0 t_0 = 2/3$ for a matter-dominated Universe and ½ for a radiation-dominated Universe.

To have a more convenient scale for H_0, astronomers use the mixed units of km/s/Mpc. A **parsec** is 3.26 light years, or 3.09×10^{13} km; a megaparsec (Mpc) is 3.09×10^{19} km. Data by Riess *et al.* (2011) indicate $H_0 = 73.8 \pm 2.4$ km/s/Mpc. The measured ages of the Universe using several methods average to $t_0 = 12.9 \pm 0.9$ Gyr (12.9×10^9 years). Because it takes 978 Gyr to travel 1 Mpc at 1 km/s, these values together give $H_0 t_0 = (73.8 \times 12.9/978) = 0.97 \pm 0.08$, which is not consistent with the relation $H_0 t_0 = 2/3$ for a critical-density Universe.

One solution to this problem would be to hypothesize that the expansion of the Universe is accelerating instead of decelerating. This hypothesis requires something that acts like antigravity on large scales, and the **cosmological constant** introduced by Einstein to cancel gravity in his early model of a static Universe could provide the required effect. But because the Universe is not static, the cosmological constant was regarded as an unnecessary complication by most cosmologists. However, in 1998, the Universe was in fact found to have an accelerating expansion, so the cosmological constant is back in a more modern guise called **dark energy**. This is a form of density that remains constant as the Universe expands, unlike matter or radiation.

The greatest difficulty in cosmology today is in determining the true distances to objects, as opposed to simply using their recessional velocities in the Hubble law. But to measure the Hubble constant, true distances as well as recessional velocities must be measured. Hubble tried this in 1929, but the distances he used were 5 to 10 times too small, and his value for H_0 was 8 times too large. For $H_0 t_0 = 1$, this gave an age for the Universe of $t_0 = 1.8$ Gyr, which was less than the well-known age of the Earth. This discrepancy motivated the development of the **steady-state** model of the Universe, in which $a(t) = \exp(H_0(t - t_0))$. The steady-state model has an accelerating expansion and a large effective cosmological constant. Because $\exp(H_0(t - t_0)) \to 0$ only for $t \to -\infty$, the steady-state model gives an infinite age for the Universe. However, the steady-state model made definite predictions about the expected number of faint radio sources, and observations made during the 1950s showed that the predictions were wrong.

The critical density is very low — only six hydrogen atoms per cubic meter for $H_0 = 74$ km/s/Mpc. A very good laboratory vacuum (10^{-13} atmospheres) has 3×10^{12} atoms per cubic meter. While the critical density is low, the apparent density of the mass contained in visible stars in galaxies, when smoothed out over all space, is at least 100 times smaller! Thus, the Universe appears to be underdense, which means that E in Equation (6) is positive and the Universe will expand forever. However, this situation is unstable. Consider what will happen as the Universe gets 10 times older. If the density is really only 1% of the critical density now, the Universe will expand at essentially constant velocity, and thus will become 10 times larger. As a result, the density will become 1,000 times smaller, since the same amount of matter is spread over 10^3 times more volume. The critical density will also change because the Hubble parameter, $H(t)$, is a function of time. When the Universe is 10 times older, the value for H will be approximately 10 times smaller. This gives a critical density that is 100 times smaller than the present density. Thus, the ratio of density to critical density becomes 0.1%. But we can start our calculations of the Universe when $t = 10^{-43}$s, and $t_0 = 10^{18}$s. If the density were 99% of the critical density at $t = 10^{-43}$s, it would be 90% of the critical density at $t = 10^{-42}$s, 50% of the critical density at $t = 10^{-41}$s, 10% of the critical density at $t = 10^{-40}$s, and so on. For the actual density to be between 10% and 200% of the critical density now, the ratio of density to critical density had to be

$$0.999999999999999999999999999999999999999999999999999999 < \rho/\rho_{crit}$$
$$< 1.000000000000000000000000000000000000000000000000000001 \quad (8)$$

at $t = 10^{-43}$s. This ratio ρ/ρ_{crit} is known as Ω, and we see that Ω has to be almost exactly 1 early in the evolution of the Universe. Figure 1.3 shows three scale factor curves computed for three slightly different densities 10^{-9}s after the Big Bang. The middle curve has the critical density of 447 sextillion g/cm³, but the upper curve is a universe that had only 1 g/cm³ of 447 sextillion g/cm³ less density and now has a density lower than the observed density of the Universe; the lower curve is a universe that had 1 g/cm³ more and is now at the "Big Crunch." To get a universe like the one we see requires either very special initial conditions or a mechanism to force the density to equal the critical density. Any physical mechanism that sets the density close

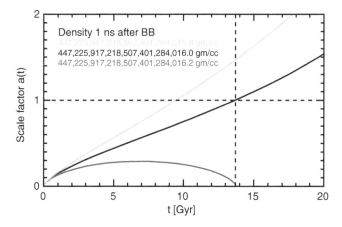

Figure 1.3. Scale factor $a(t)$ for three different values of the density of the Universe at $t = 10^{-9}$ seconds after the Big Bang. Note how a very tiny change in the density produces huge differences now.

enough to the critical density to match the present state of the Universe will probably set the actual density of the Universe to precisely equal the critical density. But most of the density in the Universe cannot be stars, planets, plasma, molecules, or atoms. Instead, most of the Universe must be made of **dark matter** that does not emit light, absorb light, scatter light, or interact with light in any of the ways that normal matter does, except by gravity.

Cosmic Background

Penzias and Wilson (1965) reported the discovery of a microwave background with a brightness at a wavelength of 7 cm equivalent to that radiated by an opaque, nonreflecting object with a temperature of 3.7 ± 1 degree Kelvin. Further observations at many wavelengths from 0.05 to 73 cm show the brightness of the sky is equivalent to the brightness of an opaque, nonreflecting object (a **blackbody**) with a temperature of $T_0 = 2.725 \pm 0.001$ K. The spectrum of the sky as a function of wavelength differs from an exact blackbody spectrum by less than ±60 parts per million. This shows that the Universe was once very nearly opaque and very nearly **isothermal** (the same temperature everywhere). By contrast, the Universe now has galaxies scattered about and separated by vast stretches of transparent space. Because the conditions necessary to produce the microwave background radiation are so different from the current conditions, we know that the Universe has evolved a great deal over its history. Since the steady-state model predicted that the Universe did not evolve, its predictions are not consistent with the observed microwave background.

Observations of the microwave background toward different parts of the sky show a small variation in the temperature, with one side of the sky being 3.36 mK (0.00336 Centigrade degrees) hotter and the opposite side of the sky being 3.36 mK colder than the average. The pattern of a hot pole and a cold pole is called a **dipole**. It is a measure of the peculiar velocity of our solar system at 369 ± 1 km/s relative to the Hubble law. The velocity is the sum of the motions caused by the revolution of the Sun around the Milky Way, the orbit of the Milky Way around the center of mass of the local group of galaxies, and the motion of the local group caused by the gravitational forces from the **Virgo Supercluster**, the **Great Attractor**, and other clumps of matter. The local group contains about 30 galaxies, of which the Milky Way and the

Andromeda nebula are the biggest; the Virgo Supercluster contains thousands of galaxies. The reader will find more details on galaxies and their clusters in Alan Dressler's Chapter 2.

After the dipole pattern is accounted for, the remaining temperature fluctuations are very small, only 11 parts per million. These tiny temperature differences were detected by NASA's Cosmic Background Explorer (COBE) satellite. This implies that the initial density fluctuations in the Universe were also very small.

The current energy density of the microwave background is quite small, as might be expected for the thermal radiation from something that is colder than liquid helium. The number density of microwave photons is 410 cm^{-3}, and the average energy per photon is 0.00063 **electron volt (eV)**. Thus, the energy density is only 0.26 eV/cm^3. This is 20,000 times less than the critical density. But when the Universe was very young (e.g., $t = 0.5 \times 10^{-6} t_0$, about 7,000 years after the Big Bang) and the scale factor was very small, $a(t) = 10^{-4}$, the number density of photons was much greater, 4.1×10^{14} cm^{-3}, and the energy per photon was also much greater, 6.3 eV. The photon density and average energy per photon correspond to a hotter blackbody with a temperature $T = T_0/a(t) = 27{,}250$ K. As the Universe expands, it also cools. Thus, the energy density of the background when the Universe was small, dense, and hot was very large, 2.6×10^{15} eV/cm^3 when $a(t) = 10^{-4}$, and dominated the density of the Universe for all times less than 50,000 years after the Big Bang.

The very small temperature fluctuations indicate corresponding density fluctuations of about 33 parts per million 10,000 years after the Big Bang. Once the energy density of background radiation becomes less than the density of matter, the fluctuations grow as the denser regions gravitationally attract more material. The process of gravitational collapse makes the fluctuations grow in proportion to the scale factor $a(t)$. Thus, in the case of a Universe with the critical density described before, the Universe was no longer radiation-dominated when $a = 0.0003$, so the fluctuations grew from 33 parts per million to 11%. This is just enough to explain the observed clustering of galaxies that we see in the Universe now. But if the Universe had no dark matter, then $a = 10^{-3}$ when the Universe stopped being radiation-dominated. Furthermore, ordinary matter interacts with light and could not move through the background radiation until the Universe was cold enough for neutral hydrogen atoms to be stable. This happened approximately 400,000 years after the Big Bang, when the temperature of the microwave background fell to 3000 K, which, coincidentally, is when $a = 10^{-3}$. Thus, if only ordinary matter had been present, the fluctuations implied by the COBE observations would have grown only to 3.3% at the current epoch. Such small density contrasts would be completely inconsistent with the far higher density contrasts we currently observe on many scales, such as galaxy clusters with amplitudes of well over 100%.

Light Element Abundances

Although the energy density of the microwave background dominated the Universe for the first 50,000 years after the Big Bang, it is even more significant during the first 3 minutes. One second after the Big Bang, the average energy of a photon was 3 million electron volts (MeV), which is a **gamma ray**. Gamma rays destroy any atomic nucleus; thus, 1 second after the Big Bang there were only protons (p or hydrogen nuclei), free neutrons (n), electrons (e), and **positrons** ($e+$), **neutrinos** (ν_e, ν_μ and ν_τ), and their antiparticle counterparts, the antineutrinos. The three types of neutrinos correspond to the three "families" of elementary particles, but, except for the neutrinos, all the particles in the second and third families (such as muons and tauons), are so heavy and unstable that they decay during the first second after the Big Bang. Weak nuclear interactions such as

$$p + e^- \leftrightarrow n + \nu_e \tag{9}$$

determined the ratio of neutrons to protons. The neutron is heavier than the proton; the neutron-to-proton ratio declines as the temperature falls. But eventually, at about 1 second after the Big Bang, the density of electrons and neutrinos falls to such a low level that reaction (9) is no longer effective. After this time, the neutron-to-proton ratio gradually falls because of the radioactive decay of the neutron,

$$n \to p + e^- + \bar{\nu}_e \tag{10}$$

which has a **half-life** of 615 seconds.

As neutrons decay, the Universe expands and grows colder. Eventually the temperature falls to the point where the simplest nucleus, the heavy hydrogen or **deuterium** nucleus (d, the nucleus having both one proton and one neutron), is stable. This occurs when the temperature is about 10^9 K, which occurs about 100 seconds after the Big Bang. At this point, the reaction

$$p + n \to d \tag{11}$$

very quickly converts all neutrons into deuterium nuclei. Once deuterium is formed, it is quickly converted into helium through a network of interactions, with the net effect

$$d + d \to {}^4\text{He}. \tag{12}$$

Because almost all neutrons that survive until T is less than 10^9 K end up bound in helium nuclei, the helium abundance in the Universe provides a measurement of the time it takes for the Universe to cool to 10^9 K. If the Universe cools rapidly, there is a large helium abundance, but slow cooling gives low helium abundance because more of the neutrons decay. The standard Big Bang model, with three types of neutrinos, predicts a helium abundance that is correct to within the 1% margin of uncertainty of current observations. Reaction (12) requires collisions between two nuclei, and if the density of atomic nuclei is low, then a fraction of the deuterium will not react. Thus, the residual fraction of deuterium in the Universe is a sensitive measure of the density of atomic nuclei. Based on the abundance of deuterium and other light **isotopes** like ^{3}He, the best estimate for the current density of nuclei of all sorts is equivalent to 1/4 hydrogen atoms per cubic meter (Copi, Schramm and Turner, 1995, Schramm, 1995). This is about 25 times less than the critical density. Because the density of the Universe must be close to the critical density to produce the observed clustering of galaxies, we find from the light element abundances that most of the mass of the Universe must be the mysterious dark matter.

In the 1940s, George Gamow and colleagues proposed that all the chemical elements were produced in the Big Bang. This proposal, described further in Virginia Trimble's Chapter 3, led to a prediction of a 5-K microwave background (Alpher and Herman, 1948), but this prediction was not followed up. The eventual discovery of the microwave background in 1964 was accidental. Why was this prediction ignored? The absence of stable nuclei with atomic weights of 5 and 8 means that the Big Bang produces only hydrogen and helium isotopes and a very small amount of lithium. When a model is supposed to produce all the elements from Z = 1 to 92, but actually only works for Z = 1, 2, and 3, its other predictions tend to be ignored. But in this case the predictions were right.

Horizons

We can see only a finite piece of the Universe. The naive estimate for how far we can see is ct_0, the speed of light times the age of the Universe. This is, in fact, the distance traveled by photons coming from the most distant visible parts of the Universe, as measured by the photons. But when one defines distances in an expanding universe, the convention is to measure all intervals at the current time, t_0. Because the Universe has expanded since $t = 0$, the earlier parts of the photon's journey get extra credit. We can compute the distance we can see in a critical density universe by dividing the age of the Universe into more and more intervals. With one interval, we get ct_0. With two intervals, we get $0.5\ ct_0/0.5^{2/3} + 0.5\ ct_0 = 1.29\ ct_0$ because the first half of the journey has expanded by the factor $1/a(t_0/2) = 1/0.5^{2/3}$. With four intervals, we obtain $1.58\ ct_0$. With a very large number of intervals, we get $3\ ct_0$ which is the distance to the **horizon**. For $t_0 \sim 13$ Gyr, this is 40 billion light years.

Consider now an observer 400,000 years after the Big Bang. The distance to the horizon is $3\ ct$, or about 1.2 million light years. This observer (really just a cloud of gas) will try to get in thermal equilibrium with the region it can see, which extends to a 1.2-million-light year radius. If thermal equilibrium can be achieved, a patch of constant temperature 1.2 million light years in radius can be created. This patch will grow to 1 billion light-years in radius as the Universe expands from 400,000 years after the Big Bang until now. But our horizon now is 40 billion light years in radius. Thus, the constant temperature patch subtends an angle of only 1/40 radian, which is only three times the diameter of the full moon. But we see an almost constant temperature over the entire sky. For a universe to be as **isotropic** (identical appearance in all directions) as the one we live in requires either very special initial conditions or a mechanism to force the temperature to be constant over the entire observable Universe.

The Big Bang model described above is in good agreement with the observed Universe, but it required very special initial conditions such as the following to explain two different facts:

1. The fact that $\rho/\rho_{crit} = \Omega$ is close to 1 today means that Ω was nearly exactly 1 initially.
2. The microwave background temperature is nearly identical in patches that could not have communicated with each other before the Universe became transparent 400,000 years after the Big Bang.

Guth (1981) proposed the **inflationary scenario,** which attempts to make these initial conditions less special (see Guth and Steinhardt, 1984). The inflationary scenario supposes that at some time during the early history of the Universe, a very large dark energy density existed, which led to a rapidly accelerating expansion of the Universe. In Russia, Starobinsky (1979) began to study Universes in which a rapidly accelerating expansion preceded the normal decelerating expansion of the Big Bang. During this inflationary epoch, the Universe was like the steady-state model, but only temporarily. A suggested prologue of the inflationary scenario is a normal Big Bang expansion until a time t_s. In Linde's (1994) model of perpetual inflation, this prologue is absent because the Universe begins during inflation. At t_s after the Big Bang, Act I, the inflationary phase, begins. During this time, the Hubble constant is $H = 0.5/t_s$. During the inflationary epoch, the Universe expands by a factor of 10^{43} or more. At about $200t_s$, the inflationary epoch ends, and Act II begins. Act II is a standard Big Bang model, but with initial conditions set during the inflationary epoch. For example, a very small patch can become isothermal, and inflation makes the small isothermal patch into a huge isothermal patch, which expands to become much bigger than the observable Universe.

But why does inflation make $\Omega = 1$ almost exactly? The answer lies in the steady-state nature of the inflationary epoch. When the Universe expands, one expects the density to go down, but in a steady-state model, the density must remain constant. Thus, there must be a continuous creation of matter during a steady-state epoch. This means that the mass M in Equation (6) gets bigger, like r^3, instead of staying constant. Then the potential energy term grows increasingly negative as the Universe inflates, in proportion to r^2. To conserve energy, the kinetic energy term $mV^2/2$ must get bigger, so V gets bigger, and the expansion accelerates as expected in a steady-state situation. As noted in the beginning of this chapter, in the first section, on the expansion of the Universe, this acceleration is equivalent to introducing Einstein's cosmological constant. But note that

$$(GMm/r)/(1/2\, mV^2) = \rho/\rho_{crit} = \Omega. \tag{13}$$

Therefore, if before inflation the Universe had $0.5\,mV^2 = 2$, and $GMm/r = 1$, so $E = 0.5\,mV^2 - GMm/r = 1$, and $\Omega = 0.5$, then after inflating by a factor of 10^{43}, we have $GMm/r = 10^{86}$. Then $0.5\,mV^2$ must be $10^{86} + 1$ to preserve $E = 1$. Thus, after inflation, $\Omega = 1 - 10^{-86}$, which is well within the tight limits given in Equation (8).

Thus, inflation solves two problems in the Big Bang model, but creates another question: why does the Universe have a large cosmological constant during the inflationary epoch? The answer to this question lies in high energy particle physics, under the topic of unified field theories. The Weinberg-Salam model that unifies the **electromagnetic** and **weak nuclear** interactions into a single **electroweak** theory requires a large **vacuum energy density**. A vacuum energy density acts just like a cosmological constant, and in the Weinberg-Salam theory, the Universe makes a **phase transition** from a state with a large cosmological constant when T is greater than 10^{15} K (or when the energy density is equivalently high) to the normal state with small or zero cosmological constant at lower temperatures. A similar unification of the **strong nuclear** force with the electroweak force gives a **grand unified theory**, or GUT. In GUTs, the transition from high-to-low cosmological constant occurs when T is greater than 10^{28} K. Either of these phase transitions could cause an inflationary epoch.

Inflation produces such a tremendous enlargement that even tiny objects such as the **quantum fluctuations** that occur on subatomic scales get blown up to be the size of the observable Universe. But while the fluctuations are being inflated, new small ones are always being created. Because the time for the Universe to double in size is constant during inflation, the power in the fluctuations in each factor of two size bins is constant. Let us measure time in units of the doubling time, and size in units of the speed of light times the doubling time. Then at $t = 10$, the fluctuations created between $t = 9$ and $t = 10$ are all about size $r = 1$ because they have existed for less than 1 doubling time. At $t = 1$, there should have been the same amount of fluctuations at size $r = 1$. But these fluctuations now have size $r = 512$ and $t = 10$. Hence, at $t = 10$, the amount of fluctuations at $r = 512$ and $r = 1$ should be the same. The same argument, applied at $t = 2, 3, 4, \ldots$, shows that amount of fluctuations at sizes $r = 256, 128, 64, \ldots$ should all be equal to the amount at $r = 512$ and $r = 1$.

These fluctuations become temperature variations, and the equality of the amount of variations on different angular scales is a prediction of the inflationary scenario. In 1992, the COBE team announced the discovery of temperature variations with a pattern that is consistent with equal

The Origin of the Universe

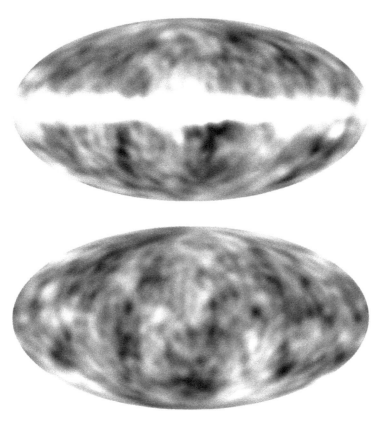

Figure 1.4. Top: The temperature fluctuations measured by the COBE DMR without subtracting the Milky Way signal. Bottom: A model sky constructed using an equal power on all scales random process.

variations in angular size bins centered at 10°, 20°, 40°, and 80°. Figure 1.4 compares a predicted sky map produced using equal power on all scales to the actual sky map measured by COBE. The two maps look quite similar, and a detailed statistical comparison shows that the equal power on all scales prediction of inflation is quite consistent with the observations.

Acoustic Scale

Observations of the CMB made since 2000 have shown a preferred angular scale of 0.8°. This scale is about 10 times smaller than the beam size of the COBE experiment. The Wilkinson Microwave Anisotropy Probe, launched by NASA in 2001, had a beam size of 0.2° and could accurately measure this preferred scale. The Planck satellite, launched by the European Space Agency (ESA) in 2009, further refined these measurements with a beam size of 0.08°. This preferred scale is related to the horizon angle discussed above. At times earlier than 400,000 years after the Big Bang, the Universe was ionized and the ionized plasma strongly scattered the photons of the CMB (**cosmic microwave background**). The pressure of the photons led to sound waves, or acoustic oscillations, that traveled at a large fraction of the speed of light. Thus, the two parts of a density perturbation will split up. The dark matter density perturbation will stay fixed, but the ionized gas perturbation will move away, traveling as a sound wave, due to the pressure of the CMB photons. Then, 400,000 years after the Big Bang, the Universe cools to the point where the plasma recombines into transparent gases. This leads to an interference pattern which enhances perturbations of a certain wavelength. This preferred wavelength fits 220 times around the circumference of the sky. This preferred spot size can be seen in Figure 1.5 (use the app at http://www.esa.int/Our_Activities/Space_Science/Planck/Planck_reveals_an_almost_perfect_Universe).

The acoustic scale can also be seen in the spatial distribution of galaxies. Galaxies are likely to form where the density is high, and for a given initial density peak that leads to a central spike of galaxies surrounded by a spherical shell of galaxies where the traveling sound wave ended up 400,000 years after the Big Bang. This separation can be measured by studying the correlation of galaxies: there is an enhanced probability that

Figure 1.5. Picture of the CMB sky seen by the ESA Planck mission.

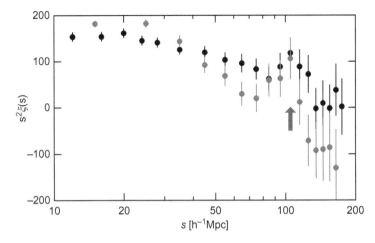

Figure 1.6. Strength of galaxy/galaxy clustering is shown on the vertical axis, versus separation, s, on the horizontal axis. The blue points are from the galaxy survey by Blake et al. (2011); the black points are from Eisenstein et al. (2005). The red arrow shows the excess probability of galaxy clustering on a separation scale of s = 142 Mpc, as predicted from sound waves crossing the Universe in its first 400,000 years.

two galaxies are separated by 142 Mpc instead of 132 or 152 Mpc. This excess probability of galaxy separations of 142 Mpc is clearly seen in the data on galaxy clustering shown in Figure 1.6. The vertical scale shows the observed strength of clustering. Size scale increases along the horizontal axis. For the current value of the Hubble Constant, $h = H_0/100 = 0.7$, the red arrow shows the increased clustering at a separation of 142 Mpc.

Current Research

Dark Matter

The small density fluctuations indicated by the small temperature differences seen by COBE can grow into the galaxies and clusters of galaxies that we see in the Universe today, but only if the action of gravity is not impeded by other interactions. The most important epoch for the growth of structures is the period just after 50,000 years after the Big Bang. At this point, the density of matter becomes larger than the density of the background radiation, which allows dense regions to collapse under the influence of their own gravity. The temperature differences measured by COBE are a direct indication of the gravitational potential differences, which are equivalent to the heights and depths of mountains and valleys on Earth. In fact, a typical gravitational potential difference corresponds to ±300 million km in a constant gravitational acceleration equal to Earth's surface gravity. But the distance between peaks and valleys in the Universe is astronomical: 300 quadrillion km. Thus, the gradient is very gentle, and only matter that moves freely downslope will be able to gather together in pools in the valleys. All chemical elements are ionized at the temperature of 30,000 K that existed 10,000 years after the Big Bang, and the resulting free electrons interact with the background radiation to produce a very strong interaction that resists the force of gravity. Thus, all ordinary matter acts like molasses and does not flow freely down the small gravitational gradients in the Universe. Therefore, most of the mass of the Universe must be made of exotic material that does not interact with radiation. It cannot scatter light, absorb light, or emit light. This is **nonbaryonic** dark matter. The nature of this dark matter is still quite uncertain.

Historically, the first candidate for nonbaryonic dark matter was the neutrino. Neutrinos are known to exist, and their number density, determined by the reactions in Equation (9), is fixed by the observed microwave background. If one of the three kinds of neutrino had a mass about 10,000 times smaller than the mass of the electron, the density of neutrinos in the Universe would be sufficient to give $\Omega = 1$. But neutrinos with this tiny mass would have a speed of about 200,000 km/s at the critical time 10,000 years after the Big Bang. Because of this rapid motion, neutrinos are called **hot dark matter**. They would thus move about 7000 light years before slowing down as the Universe expanded. The 7000 light years would increase to 70 million light years now. In any dense region smaller than this, the neutrinos would escape before the dense region could collapse, so neutrino dark matter would produce only a very large-scale structure. Our observed Universe, to the contrary, contains ample smaller-scale structures such as galaxy clusters and super-clusters, which rule out neutrinos as the principal dark matter. The simulations in Figure 1.7 show that the many small dense structures which are predicted in a CDM universe (left), are erased in a hot dark matter universe (right).

Another model for nonbaryonic dark matter assumes the existence of a new, heavy, electrically neutral, and stable particle. This particle would interact very weakly with ordinary matter and radiation, so it received the name Weakly Interacting Massive Particle (WIMP). Because such a heavy particle would be moving very slowly 10,000 years after the Big Bang, WIMPs are a form of **cold dark matter**. One possible candidate for the

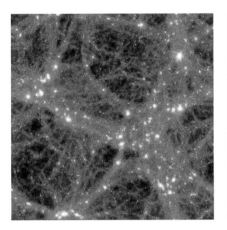 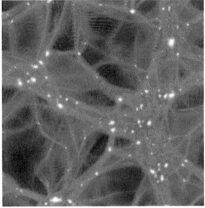

Figure 1.7. Comparison of simulations of large-scale structures formed by cold dark matter (left) and hot dark matter (right). The many small dense structures (faint yellow points) evident in a CDM Universe are smeared away by streaming in a hot dark matter-dominated Universe (such as one in which the dark matter is neutrinos). *Source*: Maccio *et al.* (2012).

dark matter is the **neutralino**, which is the lightest supersymmetric particle. Supersymmetric grand unified theories (Susy GUTs) are a currently favored class of models for the high energy particle interactions observed in particle accelerators. However, the cosmological predictions of the cold dark matter model do not depend on the nature of the cold dark matter particle. So even lacking the details, we can have some confidence in the accuracy of our general ideas about how the universe evolved, from an almost perfectly smooth state seen in the CMB at 400,000 years after the Big Bang, to the very highly structured Universe we see today.

Dark Energy

The discovery that the expansion of the Universe is accelerating (see Alex Filippenko's Chapter 4 discussion of Type Ia supernovae distances) has led to the introduction of "dark energy" to the standard cosmological model. This could be a cosmological constant as introduced by Einstein, but it could also be a vacuum energy density, like the large vacuum energy during inflation. All the data collected to date are consistent with a constant vacuum energy density, but since the large vacuum energy during inflation went away, showing that dark energy changes are possible, many scientists are trying to measure the changes in the dark energy density. Studying the dark energy density as a function of time will be a primary science goal of the ESA Euclid mission and the NASA Wide Field Infrared Survey Telescope (WFIRST) mission.

Standard Model of Cosmology

The data described in this chapter have converted cosmology from a speculative metaphysical exercise into a data-driven branch of astrophysics. Detailed calculations have been done using a standard cosmological model that has the following components: a primordial perturbation spectrum that is very close to equal power on all scales as predicted by inflation, a flat geometry or a total density equal to the critical density as predicted by inflation, and three main densities. These are ordinary matter (all the atoms in the Universe) with density $0.4189 \pm 0.0026 \; 10^{-24}$ gram/meter^3, cold dark matter with density $2.232 \pm 0.019 \; 10^{-24}$ gram/meter^3, and dark energy dominating with density 3349 ± 67 eV/centimeter^3. (When expressed in directly comparable units, the dark energy density is a bit more than twice that of the dark matter.) This model predicts a Hubble constant of 68 km/s/Mpc which is very slightly lower than the best direct measurement, and an age of the Universe of 13.8 Gyr.

Future of the Universe

Assuming the cosmic acceleration continues as it will according to most theories of Dark Energy, our observable portion of the Universe (within our horizon) will eventually encompass fewer and fewer galaxies. At some point in the very distant future (hundreds of billions of years from now), there may be no visible galaxies beyond our local group. Any newly born civilization in that era would face a far harder challenge than we do in figuring out the cosmology explained in this chapter. We may therefore be fortunate to live in an earlier part of the Universe's history when there is still ample observable evidence to show us how we got here.

Conclusion

The conclusion of this chapter is merely the introduction to the next act, the origin of galaxies, described in Chapter 2. The conditions necessary to form galaxies were established during an inflationary epoch that occurred earlier than 10^{-12} seconds after the Big Bang. Because inflation makes any preexisting structures in the Universe much too large to be observable, the temperature fluctuations observed by COBE, which were created during inflation, are the oldest structures we can ever observe.

References

Alpher, R. A., and Herman, R. 1948. Evolution of the universe. *Nature* 162: 774–775.

Blake, C. *et al.* 2011, The WiggleZ Dark Energy Survey: Testing the Cosmological Model with Baryon Acoustic Oscillations at z=0.6. Monthly Notices of the Royal Astronomical Society, 415, 2892–2909.

Brush, S. G. 1992. How cosmology became a science. *Scientific American* 267: 62–70.

Copi, C., Schramm, D., and Turner, M. 1995. Big Bang nucleosynthesis and the baryon density of the universe. *Science* 267: 192–199.

Eisenstein, D. 2015. "The Baryon Oscillation Spectroscopic Survey: Dark Energy from the World's Largest Redshift Survey". April 2015 Meeting of the American Physical Society, Abstract Z2.001.

Freedman, W. L. 1992. The expansion rate and size of the universe. *Scientific American* 267: 54–60.

Gulkis, S., Lubin, P., Meyer, S., and Silverberg, R. 1990. The cosmic background explorer. *Scientific American* 262: 132–139.

Guth, A. 1981. Inflationary Universe: A possible solution to the horizon and flatness problems. *Physical Review (D)* 23: 347–356.

Guth, A., and Steinhardt, P. 1984. The inflationary universe. *Scientific American* 250: 116–128.

Hubble, E. 1929. A relation between distance and radial velocity among extragalactic nebulae. *Proceedings of the National Academy of Sciences* 15: 168–173.

Krauss, L. 1986. Dark matter in the Universe. *Scientific American* 255: 58–68.

Linde, A. 1994. The self-reproducing inflationary Universe. *Scientific American* 271(5): 48–55.

Maccio, *et al.* 2012. "Cores in Warm Dark Haloes: A Catch 22 Problem." Monthly Notices of the Royal Astronomical Society, 424, 1105–1112.

Peebles, P. J. E., Schramm, D. N., Turner, E. L., and Kron, R. G. 1994. The evolution of the Universe. *Scientific American* 271(4): 52–57.

Penzias, A. A., and Wilson, R. W. 1965. A measurement of excess antenna temperature at 4080 Mc/s. *Astrophysical Journal* 142: 419–421.

Riess, A. G. *et al.*, 2011. A 3% solution: Determination of the Hubble Constant with the HST and WFC3. *Astrophysical Journal* 730: 119–137.

Saha, A., Labhardt, L., Schwengeler, H., Maccheto, F. D., Panagia, N., Sandage, A., and Tammann, G. A. 1994. Discovery of cepheids in IC 4182. *Astrophysical Journal* 425: 14–34.

Schramm, D. N., *The Big Bang and Other Explosions in Nuclear and Particle Astrophysics* (Singapore: World Scientific, 1995), p. 175.

Starobinsky, A. A. 1979. Relict gravitational radiation spectrum and initial state of the universe. *Journal of Experimental and Theoretical Physics Letters* 30: 682–685.

Chapter 2

The Origin and Evolution of Galaxies

Alan Dressler

Introduction

Chapter 1 discussed the **Big Bang**, the early moments of our Universe when all that we know of existed in a very different form, a swarming sea of ultrahigh-energy light and massive particles. In the beginning, this scalding bath of matter and energy was a very smooth one. Today, in contrast, the Universe is cold and dark, a darkness broken by concentrated patches of light — stars, collected into giant star systems we call galaxies. In other words, the Universe has evolved from smoothness to complex structure, an evolution that is inexorably tied to our own existence.

I should perhaps be more careful than astronomers usually are when I use the word "evolution" in regard to the Universe. "Evolution" in this context is not the evolution biologists speak of, which involves some sort of selection process, but rather evolution simply as secular change, as the *Oxford Concise Dictionary* puts it, "appearance (of events, etc.) in due succession." Whether the evolution of the Universe entails any kind of intentionality or design is, of course, beyond the scope of this text, but is fertile ground for the minds of each of us.

It is the progression of the Universe from the great symmetry and simplicity we imagine for the Big Bang, through the building of the particles we see today, with their complicated relationships, through the synthesis of the chemical elements, first in a pervasive hot plasma and then in the cores of stars, and most recently to the invention of biology and life. That is what the phrase "evolution of the Universe" means to me. It is the building of great complexity, best represented by life itself, whose essence is complex variation, from what was utter simplicity. Our own evolution is very much a part of what the Universe has been about through billions of years. We are strongly connected to the evolution of the Universe — we are not a mere sideshow or insignificant accident. This is why it is not vain or arrogant to see as the most important question: what sequence of events led to our own existence?

My chapter in this bit of storytelling is to review what we think we know and what we are still puzzled about concerning the evolution of galaxies, vast systems of billions of stars that are principal building blocks of the Universe at large. Their appearance on the scene and their subsequent evolution is a crucial plot point in the story of creation.

What is a Galaxy?

When we look into the night sky, we see stars, thousands of them, with our unaided eye, and millions if we use even a small telescope. Early in recorded history, we find a recognition that many stars are concentrated in a band that wraps around the sky, a "Milky Way" that, through the Greek words "*galaxias kyclos*," gave us our English word "galaxy". About 400 years ago, with the newly invented telescope, Galileo Galilei showed that this band is caused by the light of myriads of stars, each too faint to be seen with the eye, but together providing a fuzzy glow. The fact that these stars are collected in a fairly narrow band was correctly guessed to be an expression of the flatness of a vast disk of stars, of which our Sun was just one. More than 200 years ago, intelligent, imaginative people — one of them Immanuel Kant — speculated that there might be many other *island universes* — countless stars collected into vast disks and balls.

In the 1920s, Edwin Hubble was an astronomer at the Carnegie Institution's Mount Wilson Observatory in Pasadena, California (now Carnegie Observatories), where I myself work. Hubble put an end to centuries of debate about the reality of "other Milky Ways" when he measured a distance for the fuzzy ellipse in the sky known as the Andromeda nebula (see Figure 2.1). Hubble showed conclusively that this system lay well beyond the boundaries of our Milky Way. He did this by identifying a certain kind of star that changes in brightness in a regular way. The

The Origin and Evolution of Galaxies

Figure 2.1. Visual photograph of our neighboring "sister" spiral galaxy in Andromeda, Messier 31. The flat disk of stars is viewed at a substantial inclination angle, giving it an elliptical appearance. It contains spiral arms and dust lanes. The light of the rounder central region, the bulge, is dominated by old stars, which produce its reddish glow. The spiral arms, where the younger, hotter stars are being formed curently, are noticeably bluer. Image *Source*: Wikisky (http://wikisky.org/).

intrinsic brightness of these **Cepheid** variables had been measured by studying such stars in our Milky Way. Knowing how bright a star *actually* is allows you to calculate how far away it is, by comparing its apparent brightness to its true brightness, and employing the rule that the apparent brightness of a luminous source falls as the square of the distance.

From there, Hubble and other astronomers of the day clarified which of the fuzzy patches of light in the sky were **galaxies** and separated them from other smaller associations of star forming regions within the Milky Way, for example, the glowing patch of stars and gas that is the Orion

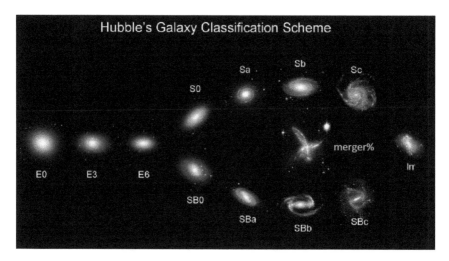

Figure 2.2. Edwin Hubble's "tuning fork" arrangement of galaxy types. Elliptical galaxies (E0–E6, increasing flatness) are round balls of stars, S0 galaxies (lenticulars) (SB0 with a bar as well as a bulge) resemble spirals with bulges, but have little or no star formation, and spiral galaxies, with bars (below) and without (above), account for more than 95% of today's big galaxies. Irregular galaxies (Irr) and mergers (center of the tuning fork) make up most of the rest. More elaborate classification schemes have been invented and utilized, but Hubble's simple scheme conveys the basic structural and star formation properties of galaxies. An important fact not conveyed in this kind of diagram is a range in size of about 10, and a range in brightness of about 100, for all these types. *Source*: University of Iowa, Department of Physics and Astronomy, NASA/Hubble Space Telescope.

Nebula (Figure 4.5). By the late 1920s, hundreds of galaxies had been identified. Hubble and others took the light of these galaxies and subjected it to spectroscopic analysis; by breaking the light into its component colors, the combined starlight that made up the vast systems could be compared with the light of our Milky Way and with that of the individual stars within the Milky Way. The most startling discovery came from the **Doppler shift** (the same effect we hear as a rise and fall in pitch of the horn from a moving car as it approaches and then passes), which revealed that many of the galaxies are receding from us at speeds of thousands of miles per second! Hubble found that the further a galaxy is from the Milky Way, the greater its speed of recession. The **expansion of the Universe**, a concept Hubble never seemed to feel completely comfortable with, became his foremost legacy. It led, eventually, to the notion of a creation event for the Universe — the Big Bang theory — and the realization that the Universe had a dynamic rather than static nature, that it had a past different from its present, and a future different still — that is, the Universe is still changing. These discoveries rank among the greatest in human history because they changed our perception of what the Universe is.

The Shapes of Galaxies

When it became clear that galaxies are other huge star systems like our own Milky Way, Hubble and his colleagues became more interested in their different forms. Galaxy shapes vary, but the variations are familial, like the difference between a giraffe and a horse, rather than between a giraffe and a jellyfish. The variations come in ranges of size and brightness, but they represent only scaled versions of what is basically the same animal. This simplicity is repeated over huge distances — a few basic shapes, a wide range in size, a lone galaxy here and there, many doubles and triples, a few groups of several large, and many smaller galaxies, perhaps one populous cluster of galaxies every so often. This is the way the Universe looks — in neighborhoods 100 million light years in diameter. (The cosmic ruler is scaled to the **light year**, the distance light travels in a year — 6 trillion miles.)

Over the decades, finding the underlying causes of this remarkable regularity became, and to some extent remains, a preoccupation of astronomical research. As is frequent in science, taxonomy was the first step — grouping subjects into classes that share some characteristics can uncover clues to the physical process that is responsible for the trait.

Some of the earliest works in galaxy classification also came from Hubble. He identified three primary types — **spiral**, **elliptical**, and lenticular (lens-shaped) — distinguished by shape and morphology rather than by size or brightness (see Figure 2.2). It is, in fact, remarkable that these different types are represented with a range of a factor of 10 in size and a factor of a hundred in brightness. Examples of spiral and elliptical galaxies are shown in Figure 2.3. Spiral galaxies like our Milky Way are the most common type; they typically appear in the sky as stretched ovals with spiral bands in their outer regions. Astronomers soon recognized that a spiral galaxy is basically a flat disk — when seen edge-on about 20 to 30 times as thin as they are across, but those viewed face-on are nearly perfect circles. Those inclined at some intermediate angle to our line of sight are seen as oval-shaped — to be precise, as ellipses.

Not all spiral galaxies are completely flat; many also swell in the middle. This *bulge* is known to be more-or-less spherical because the central regions appear roundish, regardless of the angle at which we view the galaxy. (Only a sphere looks round from every angle.) Even though the

Figure 2.3. Representative spiral and elliptical galaxies. The spiral (left) has grand spiral arms along which stars are forming in the blue knots. The "bulge" in the center, shown not only by the round spot in the middle but also by the extended red region, is the combined light of billions of older stars. The elliptical galaxy (right) is entirely made of old stars in the shape of a round ball. This particular galaxy shows thousands of old star clusters — called *globular clusters*. They show as the fine spray of points in the outer parts of the galaxy, each the combined light of about one million stars. *Source*: NASA/Hubble Space Telescope.

view of our Milky Way galaxy is obscured by our location in its dust-laden disk, we can see that it has both characteristics — the thinness of the disk is evident as the narrow band of diffused starlight that rings the night sky, and the round bulge of our galaxy shows as a prominent widening of this band toward the constellation Sagittarius. *WISE*, the Wide-Field Infrared Survey Explorer, has taken a stunning picture of our galaxy sans dust (Figure 2.4). Extending outward from the bulge is a much fainter (invisible to our eyes) spherical "halo," extending as far as the disk, that hosts many concentrated clusters of stars, called *globular clusters*, many easily visible with a small telescope. There are other characteristics of spirals, for example, some have central bar-like or lens-shaped structures instead of bulges, and the spiral pattern may be regular, on a grand scale, while others are lace-like — *flocculent* in appearance, but these differences — although important to dynamical studies (how the stars are moving) — are beyond our scope here.

Elliptical galaxies, in contrast to spirals, are simply roundish balls of billions of stars. After many years of research, it is still unclear whether

Figure 2.4. Full view of our Milky Way galaxy, taken by NASA's WISE satellite in near-infrared wavelengths. The center of this image is in the direction of the constellation of Sagittarius. That is the direction towards the center of our Galaxy. The noticeable concentration of whitish light around that direction is our galactic bulge.

they are all intrinsically egg-shaped (prolate) or onion-shaped (oblate), or whether there are some of both shapes. This has been a difficult issue to settle because we have only one view of the cosmos. We see each galaxy from only one direction. In our lifetimes — instantaneous by cosmic standards — we have no chance to travel to a place from which we might view a galaxy's "better side," and no time to let the galaxy turn beneath our gaze. Like the bulge of a spiral galaxy, an elliptical galaxy has a faint halo that extends far into space, usually spotted with large numbers of globular clusters. For many years, astronomers thought that elliptical galaxies might be a good place to begin in the study of galaxy evolution, based on the idea that their simple shapes probably meant that elliptical galaxies had the simplest histories, but the opposite turns out to be true.

Members of the S-zero (S0) class of galaxies (Hubble called them lenticulars) look like a cross between an elliptical and a thin-disk spiral. The round bulge in the middle dominates the light of the fainter disk that surrounds it, and there is no spiral pattern seen in the disk. There is also a class referred to as "irregular" that is a catchall category for the few percent that are not spiral, elliptical, or S0 galaxies. Some irregulars appear to be galaxies too small to sustain a stable, well-organized shape or pattern of star formation; nevertheless, they have much more in common with spirals than elliptical or S0 galaxies. Other irregulars seem to be galactic equivalents of train wrecks, the results of disruptive collisions and/or mergers *between* galaxies.

The spiral pattern prevalent in so many galaxies first enabled astronomers to move from taxonomy to a rudimentary understanding of the physical processes that give rise to the different types. As more was learned about the life cycles of stars, it became clear that the spiral pattern is delineated by young stars forming in thin disks — a mix of gas and older stars — and that the spiral pattern, the thinness of the disk, and the formation of new stars, are all intimately related.

Our galaxy is a typical one. It is of the most common type — a spiral — one of near-average mass, size, and luminosity. The visible part of our Milky Way galaxy, that is, the area where the stars are, is about 100,000 light years in diameter — sunlight that bounced off Earth when Neanderthals thrived on the continent of Europe has not yet crossed the breadth of our galaxy. And our Sun is a fairly typical star, though twice as massive as most. There are about 100 billion stars in the Milky Way galaxy, about a dozen stars for every human now living on Earth. Our galaxy is "rotating" in the sense that the Sun and its neighbors are moving at about 150 miles per second around the Milky Way's center. Even at this pace, it will take about 200 million years to complete one orbit — when last the Sun passed this way, dinosaurs had just begun to rule the Earth.

We are now certain of the basic picture of what is happening in galaxies. The disks of spiral galaxies are mainly made up of stars moving on near-circular orbits around the galaxy's center. Floating among the stars is a lesser mass of gas from which the stars themselves have formed. The gas is mainly hydrogen atoms, as is true all over the Universe. Helium atoms make up a significant minority, but the all-important heavy atoms, including carbon, nitrogen, oxygen, silicon, magnesium, and iron, make up less than 1%. Most of the heavy atoms are bound together in tiny grains of hydrocarbons, silicates, and ices — a thin mist of the same material of which Earth and we ourselves are made (see later chapters). The average density of the gas and dust between stars is only one atom per cubic centimeter, a better vacuum than has ever been produced on Earth, but 100,000 times denser than the gas between galaxies, and dense enough to be a starting point for star formation.

Along the spiral arms in a galaxy are thicker clouds of gas where new stars are condensing and lighting their nuclear fires. The higher density of these clouds encourages the formation of molecules of the most common elements — hydrogen, carbon, nitrogen, and oxygen, along with water and CO_2 ice, and microscopic grains of "dust" made of longer chains of molecules containing carbon and silicon, including many of the hydrocarbons found in a drop of oil on Earth (see later chapters).

In at least some cases, the spiral pattern itself seems to be the agent for star formation. A spiral arm is like a pressure wave that sweeps continually around the galaxy, the way water waves slosh back and forth in a bathtub. As it passes, the wave squeezes these **molecular clouds**, compressing their gas to a density where the pull of gravity can take over and drive an avalanche-like collapse (see Chapter 5). As each gas cloud contracts its temperature rises; this causes it to glow more brightly, and the loss of this radiant energy causes the cloud to contract further. Only fragmentation into stellar-size globules can stall this inexorable collapse: the central temperatures of these **protostars** rise to millions of degrees, unleashing the enormous power source of **nuclear fusion**. The star's central nuclear fire releases tremendous energy as heat — in gas this translates into pressure, enough pressure to counterbalance the staggering weight imposed by gravity. The star shines, for hundreds of millions, even billions of years. In its **core**, hydrogen is converted to helium in the fusion process, and in the later stages of the star's life, heavier elements are fused starting with helium as a building block. All of the elements that are heavier than hydrogen and helium were formed in the cores of stars.

This is the life of a galaxy, holding and building the raw materials for generation after generation of stars, enriching them with heavy chemical elements that in time can build planets and life.

The Ages of Stars in Galaxies

There is an obvious difference between spiral and irregular galaxies on the one hand, and elliptical and S0 galaxies on the other: star formation continues in the former, but is essentially — often completely — absent from the latter. But, the story is more complicated than this: galaxies like the Milky Way weren't "born yesterday." They are full of older stars as well. In fact, galaxies of all types have been built up through generations of stars, and all seem to have begun the process long ago. All galaxies seem to have some stars that were born when the Universe was only a couple of billion years old, but elliptical galaxies are almost exclusively made of old stars, stars that formed in the first few billion years of cosmic history. Although spiral galaxies like the Milky Way seem to have formed perhaps 20% of their stars in that same early epoch, their star formation has been continuous ever since, with a roughly constant birth rate over billions of years.

Our understanding that the history of the Universe is encoded in its stars began with pioneering observations in the 1940s by the exceptional astronomer Walter Baade, one of Hubble's colleagues at Mount Wilson. In the 1940s, Baade took very deep images on photographic plates of our nearest neighbor, the Andromeda galaxy. When Baade compared the colors and brightness of very faint stars in Andromeda with the populations of Milky Way star clusters (whose ages were reasonably well determined), Baade deduced that there are two distinct populations of stars in spiral galaxies like our own. The stars of the bulge and halo of a galaxy — the roundish part — are, on average, much older than the stars in the disk — the flat part of the galaxy. (The techniques by which a **stellar population** can be assigned an average age are nowadays *very* reliable.) Furthermore, the generally older populations have a greater representation of stars with very low abundances of heavy chemical elements, for example, carbon, nitrogen, oxygen, silicon, magnesium, and iron. As mentioned earlier and discussed in Chapter 3, these elements were formed in generations of stars — they were not produced in the Big Bang. The presence of stars with very low proportions of these elements is in keeping with the idea that these are some of the oldest stars in a galaxy, stars born before the wholesale production of heavy chemical elements. This supports the notion that the spheroidal part of a galaxy (or an elliptical galaxy, which is entirely spheroid) forms early. In contrast, the mostly younger stars in the disk, almost all of them relatively rich in heavy elements, came later.

In recent decades, astronomers have studied star formation to learn how our nearby galaxies were built. For example, they have found that the low-density halo of the Milky Way contains the oldest stars with the lowest abundances of heavy elements. These stars preserve a "fossil record" of how heavy elements were delivered by individual **supernovae** that may have induced and seeded star formation in their vicinities. They have discovered that the local star-formation rate (SFR) in one part of a galaxy is largely controlled by the density of gas available for making new stars, and have found *abundance gradients* (greater heavy element abundance in the inside of a galaxy than its outside) that suggest that galaxies are built from the center-out, with the gas required for star formation being added preferentially to the outer parts. Most recently, they have found cases of streams of stars that were peeled off (a gravitational effect called *tidal stripping*) of a smaller satellite as it plunged into the Milky Way or Andromeda. With the largest telescopes, it is possible to take spectra of these stars that prove they were part of a small galaxy that was ripped apart, by showing that the stars in the streams all share the same chemical abundance patterns and the same motions through the galaxy.

These advances came from careful study of today's galaxies: the stars in our Milky Way and its neighbors are essentially a living fossil record that tells us when they were born. It is remarkable how much has been learned about galaxy evolution from what we observe "today," but astronomers have a gift that other historians can only envy — we can observe history *as it happened*. In the last few decades, giant ground-based telescopes, and telescopes in space orbiting the Earth, have collected the light from galaxies up to 12 billion light years away, allowing us to see what the galaxies looked like when the Universe was 5 billion, 10 billion, even 12 billion years younger.

Witnessing the Origin of Different Galaxy Types

Although any large telescope can collect the light of distant galaxies and thereby look back into the past (when the Universe was *younger*), our ability to observe individual stars in the populations that Baade and his colleagues studied in nearby galaxies is very limited. In viewing galaxies that are far away, the light from their individual stars, even their star clusters, become blended together, so the study of the ages (and even more so the abundance of each chemical element) becomes very difficult. Our poorer *spatial resolution* of these faraway galaxies has been sufficient to allow us only to confirm that spiral galaxies — those galaxies with continuing star formation — remain common far back into the past, and that elliptical and S0 galaxies, though they may be less common, are also seen far back. That is, some galaxies had stopped forming stars long ago, not just in the last few billion years.

However, analyzing the light of these galaxies with a **spectrograph** to study stellar populations is not the most direct way to show how the populations of distant galaxies compare to what we see around us today. Although our ground-based telescopes have mostly been limited in spatial resolution by the turbulence in Earth's atmosphere, space telescopes can have a much clearer view. The sharpness of an image is proportional to the size of a telescope, and inversely proportional to the wavelength of light observed; for the first few decades of space telescopes, the mirrors were too small to see any details in distant galaxies. Finally, in the 1990s, the Hubble Space Telescope was launched its 94-inch mirror is large enough to show us what galaxies looked like far back in history.

When the Hubble Space Telescope was launched astronomers believed that — if it performed as promised — galaxies could be observed in some detail back to, say, half the age of the Universe. For the first couple of years of operation, the Hubble couldn't deliver on that promise, due to an error in the manufacture of its primary mirror. In 1993, a daring repair mission by Space Shuttle astronauts installed new cameras that corrected for aberrations in the original optics, restoring the extremely sharp images that the Hubble had been designed to deliver. From disappointment to exuberance, astronomers soon learned that the Hubble could reveal the appearance of galaxies further than halfway back to the beginning, far enough back in time to witness infant galaxies — within a billion years of the Big Bang.

In the years since, more than a thousand research papers have been published on what has been learned from the many thousands of hours the Hubble has spent recording images of small regions of the sky in great depth and detail. Parts of the sky that appeared completely empty in pictures taken with the largest telescopes on Earth were revealed to be crowded with extraordinary faint, distant galaxies. These special fields are called the Hubble Deep Fields, and they required pointing the Hubble for hundreds of hours, while photons from galaxies of the distant past trickled in from their 10-billion-year journey.

Probably, any astronomer would agree that the most remarkable thing about these pictures, leaving aside the extraordinary diversity and detailed information on galaxy evolution, is that the evolution can be seen directly — these are truly the pictures worth a thousand words and papers. Figure 2.5 shows individual galaxies plucked from the Hubble Deep Field images, galaxies whose distances and *lookback times* are known through measurements of their Hubble expansion velocities (the so-called velocity–distance relation that correlates how fast a galaxy is moving away from us with its distance from us — **Hubble's law**. See Chapter 1). Comparing galaxies that are seen as they were 3 billion years ago (left panel) with the Hubble types in Figure 2.2, shows great similarity, and even galaxies glimpsed 7 billion years back in time (middle panel) look more or less the same, with just a few odd cases that don't fit the normal types. However, the view 10 billion years ago is very different. The right

The Origin and Evolution of Galaxies

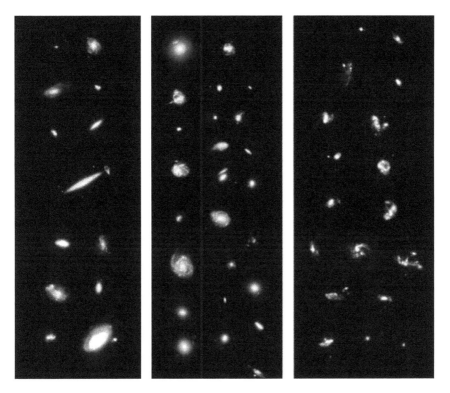

Figure 2.5. From the Hubble Deep Field images, the panels show the how galaxies are evolving through cosmic time. The left panel shows galaxies 3 billion years before the present; they match the basic types seen in the "local" Universe (today) of Figures 2.2 and 2.3. Even back 7 billion years (the middle panel), most galaxies resemble modern ones, although there are some that look somewhat "jostled." The right panel shows the galaxies 10 billion years ago, when the Universe was only 3 billion years old. Here the familiar types seem to be disrupted versions of spirals galaxies, indeed these are spiral galaxies in the act of forming stable disks of stars — *galaxies in assembly*. Note, however, that several elliptical looking galaxies are present at this early time, and that they look small, but otherwise, like the ones we have today. The basic point is that galaxy evolution can be observed *as it happened*. Source: NASA/Hubble Space Telescope.

panel shows the Universe when it was only 3 billion years old: most of the galaxies are very different looking, like spiral galaxies that have been disrupted. Indeed, astronomers believe that what we are seeing is the epoch when spiral galaxies were assembling their disks, unruly adolescent galaxies that would in a few billion years come to resemble their mature present-day descendants.

In a real sense, this may be the Hubble Space Telescope's greatest achievement, showing in pictures alone that the Universe of galaxies is evolving, that it looked different in its youth than it does today. Hubble has taken a picture worth a *million* words. Not often in science does a picture alone tell the story.

The Hubble images also show something equally fascinating about elliptical galaxies. They are not as common in the distant past as they are today, but they are definitely present in the young Universe, and unlike the spirals, they look essentially the same as they do today, smooth and round — apparently mature at a young age. Except, "pound for pound," those early elliptical galaxies were much smaller than their descendants today. This has not been easy to explain, since they would seem to have "puffed up" (the same mass, but growing in size) as time went on, and specific mechanisms that have been suggested to explain this seem to fall short of what is needed. On the positive side, such a major discovery that seems to trouble our expectations is likely the sign of something very fundamental that we don't yet understand, and that's good, because we know we're missing something fundamental about the galaxy evolution. Learning what doesn't match our preconceptions is an essential step in *understanding*.

Finally, the Hubble images of *galaxies through time* show that S0 galaxies were much less common when the Universe was very young, suggesting that most are descendants of spiral galaxies that have "run down" — stopped forming new stars. Even though they look like an elliptical galaxy with a disk added, the time sequence suggests that these are disk galaxies with a bulge (much like an elliptical) added. Again, this points us to an important conclusion: that bulges are built in spirals after they mature. As described in the following section, the swallowing up of smaller satellite galaxies by larger galaxies, referred to as minor mergers, could be the solution to both problems, how an elliptical galaxy can puff up without gaining much mass, and how a bulge can grow in a predominantly disk galaxy.

Galaxies and Giant Black Holes

In addition to stars, gas, and dust, elliptical galaxies — and disk galaxies with prominent bulges — harbor giant black holes in their holes in their centers. These massive mysteries of general relativity hide millions to billions of solar masses in a space where…well actually, in no "space" at all. Clues to this extraordinary fact of nature came first in the 1960s, when newly developed *radio telescopes* tuned into the sky — gathering a form of light that is less energetic than the visible light we are accustomed to, but light nonetheless. Radio astronomers were scanning the sky and pointing at astronomical sources of all kinds — Milky Way stars, clouds of gas and dust, and of course other galaxies, when something astonishing was discovered. Most galaxies were found to "glow" with weak radio emission from starforming gas clouds, but a tiny fraction had powerful radio beacons at their center. As radio telescopes grew in size and sensitivity, they produced radio pictures that showed remarkable jet-like emission emerging on opposing sides of the galaxy's center. Figure 2.6 (left-top) shows

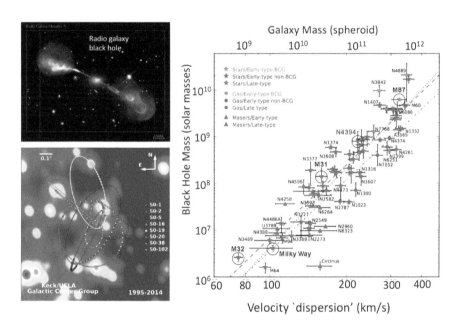

Figure 2.6. Supermassive black holes. The top-left image is a composite of an optical light picture of an elliptical galaxy (shown in white), superposed on a radio telescope "image" showing radio emission from two jets of electrons expelled from the galaxy's center (pink). Identification of the first "radio galaxies" led to the discovery of supermassive black holes at the centers of elliptical galaxies and bulge-dominated disk galaxies. These are the most powerful long-lived energy sources in the Universe. On the right is a 2013 compilation of the mass of black holes in galaxy centers compared to the mass of the galaxy's "spheroid" of stars (top scale), and to the range (or spread) of the speeds of the stars in the spheroid (the velocity "dispersion" — bottom scale). The author proposed this relationship based on the data for five galaxies with supermassive black holes, including the Milky Way's nearest neighbors, Messier 31 (Andromeda) and its companion elliptical, Messier 32. These early observations of M31 and M32 were the first to suggest that supermassive black holes are common to galaxies, and that their formation and growth is related to the formation and growth of the entire galaxy. The bottom-left picture shows the actual orbital paths of stars around the Milky Way's 4 million solar mass black hole, from the UCLA Galactic Center Group, led by Andrea Ghez. *Source*: National Radio Astronomy Observatory.

one famous example, a radio galaxy firing what we now know are beams of high-energy electrons, traveling at near the speed of light to far beyond the extent of the galaxy of stars.

Optical telescopes were soon trained on these radio galaxies. Most resembled giant elliptical galaxies, but in visible light failed to show anything unusual at the galaxy center. A few did, however, and these turned out crucial to understanding what was going on. Rather than finding galaxies at the positions of the radio sources, the optical images showed what appeared to be a star, and stars were not expected to be strong sources of radio waves. But, when spectra were taken using the giant 200-inch. Palomar telescope by astronomers at the Hale Observatories, it was realized that these were not stars in our Milky Way, but — incredibly — objects of the distant Universe. They were dubbed quasi-stellar radio sources (false stars), or **quasars**. Caltech astronomer Maarten Schmidt's identification in 1963 of common hydrogen emission lines in quasars proved that they were very distant objects, billions of light years away, brighter than the brightest galaxies that could be seen at that great distance. To appear so bright these sources needed a huge source of energy, and that much energy could not be coming from stars powered by nuclear fusion. Some exotic and unknown source of energy was needed.

In 1969, Cambridge astrophysicist Donald Lynden-Bell suggested that huge black holes could be the answer. In the 1960s, black holes were still only theoretical predictions of Einstein's general theory of relativity. However, the fact that some stars appeared to be orbiting massive, but invisible objects was indirect evidence that stars 10-times or more as massive as the Sun could have collapsed into black holes. Lynden-Bell calculated that a quasar *"engine"* required a black hole with the mass of millions, even billions of suns. It may seem paradoxical that a huge black hole could power a quasar, since black holes are places in space-time from which nothing can escape — not even light. However, Lynden-Bell imagined a whirlpool-like disk of gas feeding the black hole, releasing huge amounts of potential energy as the gas violently pushed its way in. This very hot **accretion** disk outside but around the black hole would radiate as much energy as the combined light of all the galaxy's stars (see Chapter 4).

Five decades after Lynden-Bell proposed that feasting super-massive black holes are the engines of quasars, all the pieces of his picture have been confirmed. The final nail came with the actual detection in the centers of nearby galaxies of huge dark masses — confined to a volume tens

Figure 2.7. Optical image of Messier 87, the giant galaxy in the Virgo galaxy cluster. The fuzzy white light is from hundreds of billions of old stars. However, the unusual long lumpy blue "jet" emerging from the right side of the galaxy's center is a beam of high-energy electrons that must have been accelerated by powerful forces near the giant black hole in the galactic center. *Source*: NASA/Hubble Space Telescope.

of light years across. The supergalaxy Messier 87 in the Virgo cluster was the first persuasive example: M87 has a jet of high-energy electrons that can be easily seen in optical pictures of the galaxy (Figure 2.7). Caltech astronomers Wallace Sargent and Peter Young obtained spectra that showed the speeding up of the stars at the galaxy's center: the gravitational pull of a mass of several billion suns concentrated at the very center was the best explanation. Although M87 is not a quasar, its jet provided additional circumstantial evidence for a supermassive black hole.

I followed Sargent and Young's work in the late 1980s with observations of our nearest neighbor, the Andromeda galaxy, and its small companion — the dwarf elliptical galaxy Messier 32. For both galaxies I found stars speeding up dramatically at their centers, something hard to explain without massive black holes, a few million **solar masses** for M32 and about 100 million for Andromeda. Particularly exciting was that Andromeda was the closest big galaxy to the Milky Way, and neither of these galaxies showed any quasar-like behavior. I took this to mean that all big galaxies had massive black holes at their center. Even if they weren't spewing massive amounts of energy like a quasar, they could have been in the past, and might be in the future, when they were again *feeding* as Lynden-Bell had proposed.

From these few examples, it seemed that the mass of the black hole scaled with the mass of the spheroidal component of stars. This spheroid is everything for M87 and M32 but only about 30% of Andromeda's mass, which has a more massive disk. Considering that our Milky Way has only a small bulge, I predicted that its black hole had to be less than a few million solar masses, and added another black hole detection by Canadian astronomer John Kormendy for the bulge-dominated spiral galaxy Messier 104. From these first five galaxies with measured supermassive black holes, I suggested that the black hole mass increased in proportion to the mass of stars in the spheroid or bulge, and not the mass of the whole galaxy. The supermassive black hole in each galaxy had a mass of around 0.5% of the mass of stars in the galaxy's spheroid.

Gratifyingly, this relationship has been confirmed and has been considered one of the more important discoveries in this field. Today's data confirm how strong is this correlation (Figure 2.6, right), using measurements of black hole mass (now made in several complementary ways) for many dozens of galaxies. The relationship between these two very different processes, the one that produces the giant black hole, and the one that builds the enormously bigger galaxy, is considered as surprising evidence that a galaxy's black hole grew with the galaxy itself. However, even three decades later the processes involved are poorly understood.

The detection of the supermassive black hole in our own Milky Way — in the end — provided the crucial evidence that convinced the few remaining skeptics that big black holes are in the centers of all galaxies, at least the ellipticals and the spirals and S0 galaxies with bulges. For the single case of our Milky Way galaxy, it is possible to measure the motions of individual stars as they orbit the black hole, leading to a very accurate measurement of 4 million solar masses. Figure 2.6 (lower left) shows the data from UCLA astronomer Andrea Ghez and her collaborators, who have been following these stars for many years as their orbits around the black hole progress. Nothing else known to physics could pack so much mass into a volume only light years across. Each year, as these stars proceed precisely around their predicted orbits, we confirm that the Milky Way's big black hole is real — evidence that all the others are as well.

How Did Galaxies Form? — The Basics

There is a simple explanation why there are galaxies in the Universe: gravity is responsible. Gravity is a characteristic of matter — a universal attraction between any and all mass, in direct proportion to the amount of mass and in inverse proportion to the distance of separation — squared. We cannot say why there is gravity and do not yet understand fully how the force of gravity is exerted, but neither its reality, nor is its role in building galaxies, is in question.

We can see back to a time when the Universe had no galaxies, when all matter was very smoothly distributed. In 1989, a NASA satellite called the Cosmic Background Explorer (COBE), recorded a remarkable picture of the Universe in *microwave light* that showed in unprecedented detail how the Universe appeared 400,000 years after the Big Bang (see Figure 1.5 for a higher-resolution view of this microwave background map). At this point in time, the Universe had a temperature of around 3000°C, far too hot for stars or galaxies to form. However, physicists had predicted very small undulations in the density of the gas that would grow through gravity into the modern Universe of a large-scale structure made up of galaxies like the Milky Way. With far greater sensitivity than had ever been possible, *COBE* discovered these fluctuations in the form of tiny temperature variations, at a level of 1/1000 of 1% over adjacent parts of the sky. Although we are not yet sure, even decades later, where these *fluctuations in density* came from, there is growing consensus that they were nothing more — and nothing less — than inevitable quantum fluctuations from the earliest moments of the Universe, amplified enormously during a period of extraordinary exponential growth in the size of the Universe — the **inflationary** epoch (see Chapter 1). The pattern is not expected to be regular, like a checkerboard or herringbone, but more like the chaotic pattern of waves in the middle of the ocean — some large, more medium-sized, and many small. Using subsequent, even more sensitive microwave space telescopes in the last two decades, we have even measured the relative proportions of different size waves in this early sea of matter — a *power law* (approximately $1/f$, where f represents the *frequency* of occurrence), a form that is very common in nature, for example, in the number of peaks of different heights in a mountain range. In this type of distribution, how much of something there is varies inversely with the scale size (something twice as large is twice as rare).

With this as prelude, we know that, in (truly) broad brush, what happened in the first billion years of the Universe's 13.7-billion-year history is that gravity collected amounts of matter into vast "continents" separated by a comparatively empty sea. Within these continents, gravity has further drawn matter into denser clouds of gas that fragmented and collapsed into stars, at last stabilized against the inexorable pull of gravity by the enormous release of energy from nuclear fusion in stellar cores.

Galaxies formed because of gravity. But why are there variations in form? Why are some galaxies full of young stars outlining spiral arms while others — the elliptical galaxies — contain only old stars wending their way to eventual death? (See Figure 2.3.) Now that our telescopes are powerful enough to look back billions of years in the past, we can see a connection between the galaxy types and their history: the elliptical galaxies we see inhabited the regions of space more densely populated with galaxies, where galaxy formation began earlier, and vigorous star formation led to the more massive galaxies we see today. In contrast, spiral galaxies are still forming stars today; they are generally less massive systems that reached their peak of star formation when the Universe was 5–10 billion years old.

What about the distribution of galaxies themselves? We have learned, especially during the past decades, that galaxies are not strewn randomly through space, but are distributed in lacy patterns and dense clusters, separated by nearly empty voids — more like sculpture than spatter. Surely gravity is responsible for this too, but what is it about gravity working on the initial conditions of the Universe (that is, the physics of the Big Bang) that carves these almost deliberate-looking patterns? Spiral galaxies are more commonly found in low-density regions; elliptical galaxies are found more commonly in denser groups, and most commonly in very dense clusters. Why? What mechanism regulates which kinds of galaxies form, and why do their relative numbers vary in different space environments?

The 1970s began a huge breakthrough in our understanding of why the modern Universe exhibits these patterns made up of galaxies, how this "large-scale structure" grew from a Universe that was originally remarkably *smooth*, and how this relates to the formation of galaxies themselves. Astronomers collected the first conclusive evidence that there is **dark matter** in the Universe — matter that does not emit light — that is responsible for the large-scale structure and the gathering together of ordinary matter to make stars and galaxies. The discovery that stars in spiral galaxies orbiting far from the galaxy's center travel at speeds well in excess of what the collective gravity of the galaxy's stars would generate, as if some much greater gravitational force is provided by matter we cannot see, was at first hard to accept. Today, we have abundant evidence, especially the bending of light predicted by Einstein's general theory of relativity that produces arc-like images of extremely distant galaxies as their light crosses a massive cluster of galaxies.

As described in Chapter 1, modern measurements indicate that the Universe contains about six times as much dark matter as the normal matter — mainly protons, neutrons, and electrons — that we are familiar with. The big difference is that the electromagnetic force is the chief agent of interaction for normal matter particles, while gravity seems to be the only force that binds dark matter together. Thanks to extraordinarily powerful modern computers with large numbers of processors, it is now possible to make a remarkably detailed model of how gravity and universal expansion caused an initially smooth distribution of dark and normal matter to evolve into the clumpy, frothy distribution of matter we see today. The pull of gravity resists the expansion of the Universe in regions of slightly higher density; over time the dense regions grow and the sparse regions empty, leaving voids in the distribution of galaxies. It is this competition between the Universe's expansion and local contraction via gravity that produces the filamentary patterns of large-scale structure.

Normal matter is essentially entrained with the dark matter through the latter's dominating gravity: if that were all there was to it — clumps, voids, and filaments — a very dull Universe that is globally expanding and locally contracting into oblivion would be the end of the story. But, the normal component of matter interacts strongly, through electromagnetism: it is the binding together of protons, electrons, and neutrons into atoms that leads to the indescribable complexity of galaxies, stars, planets, and life. Unlike dark matter, the electromagnetism of normal matter causes light — photons — to carry away kinetic energy from a gas, and this results in contraction to a higher density. A runaway process of this sort happened everywhere in the young Universe where the matter density was sufficiently high, resulting in a studding of the dark matter filaments with concentrations of normal gas of steadily increasing density — these were embryonic galaxies. The first stars formed in these

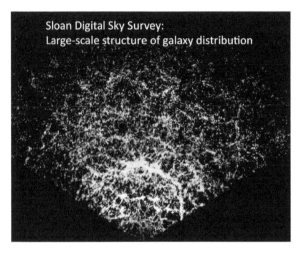 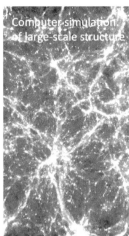

Figure 2.8. The lacy patterns of knots and filaments shows in this map from the Sloan Digital Sky Survey, where each galaxy in a slice of the Universe is represented as a point. Our Milky Way galaxy sits at the bottom point of the pie-slice, and the increasing velocity of the galaxies away from us — the Hubble expansion of the Universe — is shown by the progression of colors from red close to the Milky Way to blue at the top (the opposite of the normal convention, which would mark increasing *redshift* with a color going from blue to red). On the right, the ambitious *Millennium* computer simulation (Volker Springel and collaborators) shows that an expanding Universe shaped by the gravitation of dark matter produces just such patterns, including "great walls" dense with galaxies and empty "voids" nearly empty of galaxies — the result of expansion on a global scale and contraction by gravity on a more local scale.

precipitations of dense globs of normal matter. Their nuclear fires, stoked by the fusing of atomic nuclei, injected energy back into the gas, stopping their relentless contraction and producing stable sources of heat and light for billions of years to come (see Chapter 5).

State-of-the-art computer models (such as the Millennium numerical simulation shown in Figure 2.8 from German astrophysicist Volker Springel and his collaborators) can in fact replicate a Universe that looks essentially identical to ours in terms of large-scale structure — a monumental achievement. This theoretical approach has had considerably less success illuminating how galaxies themselves form, how they get their different shapes, and what regulates (and ends) starbirth through the eons to come. The problem is that, while gravity can be modeled with fidelity, the interactions of atoms through electromagnetism require computer simulations of far greater finesse and *dynamic range*. Here, dynamic range refers to the difference between the largest size structures — for example, a galaxy — and the smallest — for example, a single star in the gas cloud of its birth — that must be simultaneously modeled by the computer code that applies physical processes. For this reason, *predictive* models of how galaxies might evolve to their variety of forms, or what regulates or curtails starbirth, are still beyond our capabilities, although continual progress is being made toward that end. For now, the models we have that merely reproduce existing observations are insufficient; the ability to predict behavior in as yet unobserved situations is the essential component of a scientific theory.

There is an instructive example of the difficulty facing theoretical simulations of the Universe in predicting the details of galaxy formation. The picture described above, of a Universe whose evolution is driven by the gravitational interactions of dark matter, makes a clear prediction that the buildup of structures, both large-scale patterns and individual galaxies, will proceed in *hierarchical* fashion. That is, small structures form first, and bigger structures assemble over time from the smaller ones. This is clearly the case for dark matter, and it seems to be well born out in the observed patterns of large-scale structure: they match very well in scale and form what the computer-made numerical simulations predict. However, theoretical predictions that galaxies would also form hierarchically are in stark conflict with observations of both nearby and distant galaxies. It has been known since the 1980s that most massive galaxies are the oldest in terms of their populations of stars, not the youngest. In other words, the formation of galaxies appears anti-hierarchical. Theoretical models were "modified" to achieve agreement with the observed Universe, but these additions to the theory were more about behavior (if this, then do that) than they were based on real physics. Only in the last decade has it become clear why the computer simulations failed: real galaxies host outflows of gas driven by the energy released in star formation, and inflows of gas back into the galaxy. The competition of inflow and outflow regulates the growth of the galaxy by the rate of star formation — astronomers call this *feedback* — in a complex way that our observations are just beginning to reveal. Feedback is why the growth of galaxies appears to be largely anti-hierarchical, with the largest galaxies having their star formation shut down earliest. Unfortunately, it is not yet possible to introduce algorithms into the numerical simulations that adequately describe the physics of feedback, so the theoretical models have not predicted anything that we have not learned already from the observations.

For two decades, numerical simulations of galaxies without these processes stressed the role of mergers between smaller galaxies as the principal way by which galaxies grow. Mergers were credited with most of the important features of galaxy evolution, including the difference between spiral and elliptical galaxies: the latter were explained as cases of spirals that crashed together and merged, ending star formation and destroying the disks that had been the principal feature of the spirals. Observers consistently pointed out the problems with this model *vis-à-vis* observations, chiefly that the stellar populations of elliptical galaxies, in age and heavy element abundances, do not match the *ingredients* provided by two spiral galaxies.

The most recent numerical simulations, now fitted out to replicate (but still not predict!) the effects of feedback that regulate galaxy growth, no longer identify mergers as the principal agent of galaxy growth. Instead, it seems that only the relatively rare massive galaxies, about 3 times as large as the Milky Way and up, reached that great mass through mergers of spheroidally dominated galaxies, systems that were largely finished with star formation. In contrast, most galaxies grew to their present size through star formation — as we will now see — when we look back in cosmic history. Instead, the role of mergers now focuses on large galaxies absorbing smaller satellite galaxies — called *minor mergers*, which can affect both the growth of the largest galaxies (particularly their sizes, as discussed in the following), and the way spiral galaxies build large spheroidal "bulges" of their own.

The Star Formation Histories of Galaxies

As was described earlier, the many generations of stars in a galaxy form a kind of fossil record of how the galaxy formed, how it grew, and how it matured. This is a *living fossil record*, since most stars continue to shine for the full 13.7 billion years age of the universe. However, stars form in clusters according to a *power law* distribution in their mass. As discussed further in Chapter 5, this means that a few very massive stars (say, around 20 times as massive as the Sun) are born along with a few dozen massive stars (3–10 solar masses), a few hundred stars about as massive as the Sun, and fully thousands of stars half as massive as the sun — the most common kind of star. In fact, astronomers have found this kind the *power-law distribution function* to be *very* common in the universe — two other examples are star and galaxy clusters, with few bright (or massive) stars or galaxies, many more not-so-bright (or not-so-massive), and a far larger number that are faint (or low-mass).

The different lifetimes of different mass stars make it possible to determine when a star cluster was born — its age. The more massive the star, the shorter its lifetime, because the energy emitted depends strongly on a star's mass: even though there is more fuel for the nuclear furnace, a very massive star uses up its supply very quickly, relatively speaking. The lifespan of a star ten times as massive as our Sun is about ten million years, compared to the Sun's ten-*billion*-year lifespan. To learn the age of a star cluster, we observe which stars are still alive and which have ended their lifetimes and essentially disappeared, by exploding as supernovae (in many cases leaving behind a super-dense "neutron star"), or collapsing to very dense and faint **white dwarf** star (see Chapter 4). By establishing where the break point is between stars that are still shining and those that have evolved to the next phase (a giant star), we can determine how long it has been since that star cluster was born.

You might think, then, that we could quite readily determine the past birth rate of stars in a galaxy, as simply as — for example — the birth rate of people can be reconstructed from the age distribution of a stable, closed population. However, there are obvious limits: to find out the birth rate in the United States since 1900 you would have to count the many people who are no longer with us from the first half of the 20th century, in addition to the age distribution of those still alive. The same is true for a galaxy: the young star clusters we see may well tell us the stellar birth rate (a.k.a star formation rate — *SFR*) for the last hundred million years, but most of the star clusters born a billion years or more ago have dissipated, their stars wandering away to join the other stars in the galaxy's disk. Unfortunately, it is as difficult to determine the ages of *individual* stars as it was *easy* to find their age when they were born together in a cluster. This makes recovering the *star formation history* in a galaxy difficult but possible for the nearest few dozen galaxies, where individual star clusters can be identified. For the far numerous distant galaxies, it is impossible to resolve individual stars and clusters.

Nevertheless, astronomers are beginning to make progress in describing the different star formation histories of galaxies over their full lifetimes, using other indirect but still powerful methods. A huge step forward was the recognition that the star formation history for a representative *collection* of galaxies could be determined by putting all their data together. Observations of galaxies at earlier times record the birth rate of stars at that time — an instantaneous measure — through the light of the brightest, hottest stars forming at that time. In this way, the *total* star formation rate for an ensemble of galaxies could be measured at every time in the past — astronomers like the term *epoch*. Thus the total star formation history of a representative volume of the universe can be calculated. It is important to note that this will be a view of star formation on a vast *global* scale. We will not attempt here to examine the complex fine details of how an individual star forms — those specific processes will be covered in Chapter 5.

This inspired approach was developed over a 20-year period through the efforts of a number of leading researchers. In particular, astrophysicist Piero Madau wanted to know, essentially, the star formation history of the whole universe (rather than galaxy by galaxy) to compare with numerical simulations of the kind described earlier. Figure 2.9 shows a contemporary version of "the Madau diagram." The horizontal axis covers nearly all of

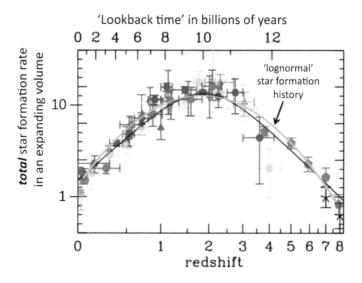

Figure 2.9. The "Madau diagram" plots the global star formation rate (SFR) over the history of the universe. Rather than focus on individual galaxies, the total rate of starbirth for a representative volume of the universe is followed over the full range of cosmic time (the horizontal axis). The vertical axis records the combined SFR for all galaxies within a representative volume. The SFR rose rapidly in the first 2 billion years, increasing by a factor of 50. After reaching a broad plateau 3–4 billion years later, the global SFR began a decline, a factor of 10 from the peak to the present epoch.

cosmic time, from the appearance of the first galaxies (the right) to the present (the left), while the vertical axis shows this cosmic SFR on an arbitrary scale. Individual points come from dozens of measurements of the SFR for samples of galaxies at different epochs. The points scatter from a smooth curve because of the difficulty of making such measurements, particularly for very distant galaxies, and also because of *cosmic variance* — the statistical variations caused by the non-uniform filling of space — the voids and filaments of large-scale structure discussed earlier.

The principal features of Figure 2.9 are (1) a dramatic rise in the SFR (an increase of 50 times) in the first two billion years of cosmic history (starting at a lookback time of about 13 billion years), (2) a leveling off to a broad peak when the universe reached an age of about 4 billion years, and (3) a plunge in the SFR from that time to our own. This last feature is to me still the most surprising: early in my career, the common lore (based on scant information, to be sure) was that star formation in the universe had been rapid at the very beginning, sharply declining soon thereafter, and slow and nearly constant for the last 5 billion years. We now know we live near the end of star formation: 10 billion years from now, star formation will have been reduced to a trickle and the brightly lit spiral arms of today's galaxies will be faded and dull.

Why this behavior? What regulated the early growth and what is responsible for the present crash? Answering that question requires a return to the star formation histories of individual galaxies. If we can compare the history of individuals to this global trend, we should start to learn why the Madau diagram looks as it does.

At present, there are two basic schools of thought about this. One is that all galaxies follow a similar route: rapidly growing at early times (when there was plenty of gas — the raw material of star formation), then slowing, peaking, and eventually declining. In this model star formation is shut down for each galaxy in a process called *quenching*. The alternative is that each galaxy follows its own growth pattern at a pace determined at birth: the denser the material from which the galaxy was formed, the faster the internal "clock" that dictates the rise and fall of its star formation history.

It would seem we could choose between these two models by looking in the past and watching one or the other, or both, in action. But the changes we're looking for take at least 100 million years, if not billions, so it is impossible to watch any given galaxy change in our lifetimes. We find ourselves in the situation of paleontologists who must assemble the evolution of species from snapshots taken at different times — the fossil record. Our fossils — galaxies and stars — are still very much "alive" as we look back in time to study them, but they are — like fossils on Earth — frozen moments in their lifetimes.

To start our analysis, it is fairly easy to sort the bulk of modern galaxies into two broad groups, based on the optical colors of their starlight. The galaxies with the reddest colors (elliptical and S0 galaxies) have not formed any stars for billions of years. Long ago, before they had exhausted their supply of interstellar gas, they must have had many recently formed massive hot stars on the main sequence, but nearly all of these have now finished their lives. What we see today is only the cooler stars, which will survive on the main sequence for tens of billions of years, and the red giants produced by the intermediate-mass stars which have recently completed their main sequence evolution. The only evolution occurring in these red galaxies is "passive" fading as the entire population ages. First the bluest stars die, and then progressively cooler and cooler stars end their main sequence lives, leaving behind a "red and dead" galaxy. Currently, **all** of the massive galaxies, up to a trillion times the mass of the Sun, are in this red category — their star formation is essentially finished.

On the other side of the Hubble tuning fork diagram (the right side of Figure 2.2), are the galaxies which still retain gas which continues to form stars today. This ongoing star-formation is most widely found in the flattened spiral disks. The surfaces of the youngest massive stars are much hotter than the Sun's, giving the entire star-forming galaxy a blue color. Galaxies with masses around that of our Milky Way (a hundred billion Suns) or less, can be either passive and red or star-forming and blue, with the latter becoming more common in the smaller galaxies.

Now our task, like that of paleontologists, is to look for *missing links* between the passive red galaxies and the blue star-forming galaxies; these can help us choose between the model of rapid quenching and the model of slow, steady decline. If quenching is the common way that galaxies transform from star-forming to passive, then we should catch galaxies in the act of quenching, or soon after. Shortly after it has formed its last stars, the starlight of a quenching galaxy should, in a hundred million years or so, change from blue to red. Indeed, there is a known population of galaxies with such intermediate colors, indicating a mixture of young and old stars. One interpretation was that galaxies in-between the blue and red colors are *quenching* — moving rapidly to redder colors after losing their ability to form stars.

However, observations show that very few of these intermediate galaxies are actually quenching. My colleague Louis Abramson and I found that most of them have not terminated star formation. They are not rapidly transforming into red and dead galaxies, but will probably maintain their intermediate appearance for a long time. Their intermediate colors result from the combination of a very old "passive" bulge plus a blue disk that continues to convert interstellar gas into new stars, just as in normal blue spiral galaxies. How red or blue the galaxy depends on the relative proportions of stars in the bulge or the disk. Abramson and I also used spectra from the Sloan Survey of these intermediate-color galaxies to show that only *a few percent* are genuine cases of rapid quenching in which star formation recently ended.

The work just described was part of a larger team effort that included Gus Oemler, Mike Gladders, Bianca Poggianti, and Benedetta Vulcani. Our aim was to explore the more general question of the diversity of histories in star formation for galaxies. Our group made spectral observations to measure the star formation rate and masses of galaxies over lookback times as large as 6 billion years. We began with the common practice of describing star formation histories as falling (with an *exponential* decline) from some very early time, say, a few billion years after the Big Bang. This falling exponential model predicted that there should be no galaxies whose SFR *increased* over the last 6 billion years, similar to what we see for the "average of all galaxies" in the Madau diagram's decline over this period. However, using the data from our study, Gus Oemler was the first to realize that this inference was wrong: many individual galaxies did have *rising* rates of star formation during the epoch covering 3 to 6 billion years ago.

To find a replacement for the "exponential decline" model of star formation histories, Mike Gladders suggested the rising-falling form observed in the Madau diagram. He reasoned that *individual* galaxies might have the same form, with different parameter values for each galaxy: the time of the peak of star formation, and the width or duration of the star formation in time. Gladders wrote a sophisticated computer code to determine these two parameters for each of the thousands of galaxies in our data sample, based on two *constraints*: reproducing the star formation rates and masses at each epoch, and requiring that the sum total of all the star formation histories match the Madau diagram. Figure 2.10 shows a realization of the calculations. Each individual line shows the model for star formation history of an individual galaxy.

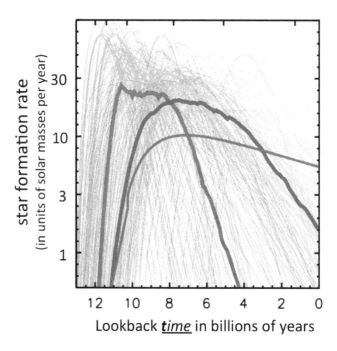

Figure 2.10. Star formation histories. The panel shows individual modelled SFHs — one line per galaxy. The orange lines are today's "passive" galaxies, while the blue lines represent galaxies which are still forming stars. The heavy red and blue lines are averages of today's passive and star-forming galaxies. The magenta line represents the quenching model, where each galaxy follows the same SFH until star formation abruptly ends.

This exercise confirmed that star formation in today's passive galaxies peaked sharply early in cosmic history, and then finished long ago. Other galaxies are still forming stars today. They cover a broad range of peak times and the widths of their SFHs are systematically larger than the passive galaxies, as they must be to continue star formation today. Although there are only a tiny fraction of *rising* star formation histories in today's universe, 20% of the galaxies seen at 6 billion years of lookback time — the earliest in our sample — have *rising* star formation rates at that time. The percentage of "risers" likely increases further at even earlier times.

Our model requires no distinct *quenching* to match the observations: most galaxies have long and continuous star formation that eventually declines, most substantially. The model still requires an explanation of why a peak is reached and then declines significantly from that time.

The Origin and Evolution of Galaxies

There is general agreement that the "food supply" — the gas from which stars and galaxies are built, is still plentiful around galaxies in decline, but some processes appear to heat and/or alter the gas properties that make this fuel supply unusable for star formation. Accreting black holes are one suggested explanation, another is the energy released by star formation, particularly supernovae, but these are early days for understanding what regulates star formation, and why star formation histories are very diverse.

The Frontier of Galaxy Evolution

The earlier we probe in cosmic time, the more the light from these ancient ancestors shifts to the red. The first- and second-generation cameras on the Hubble Space Telescope, sensitive mostly to visible light, were barely able to cross the billion-year-old horizon, that is, to see infant galaxies within one billion years of the Big Bang. However, the Wide-Field-Camera 3, installed in 2009 in the final Space Shuttle visit to the Hubble, carried new detectors more sensitive to infrared light; thus pushing the frontier back into the first billion years of cosmic history. The search for galaxies in the early Universe requires extremely deep multicolor images in order to discover very faint, slightly fuzzy blobs that show up only in infrared filters. What makes this technique so powerful is that the visible light of the galaxies from the first billion years was completely absorbed by the hydrogen gas in and around these galaxies: what we see as visible light coming from these ultra-distant objects is actually redshifted ultraviolet (UV) light leaving the galaxy. For galaxies younger than a billion years old, each photon is so energetic that it is easily absorbed by a hydrogen atom.

Figure 2.11 shows examples from the early years of Hubble-WFC3 observations, images of galaxies whose **redshift** implies a lookback time of about 13.2 billion years (a redshift of 8), only 500 million years after the Big Bang, when the Universe was 9 times smaller in each dimension than today. Compared to other galaxies in the *foreground* (which in this case is most of the Universe), these infant galaxies are incredibly faint, visible only in infrared images that required about 100 hours of light collection. The tiny spots of light from these proto-galaxies are greatly enlarged (in the boxes) for seven examples — the top four images show only a single dot-like image, but the three bottom images show extended light and, for one, multiple images. All of the light is coming from regions with intense star formation, regions that are about 10% the size

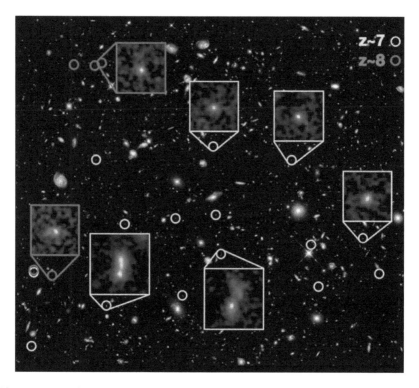

Figure 2.11. Galaxies in the Hubble Deep field at lookback times of about 13.0 billion years, 700 million years after the Big Bang. The blue color of each image recalls the fact that the light coming from these young star systems is dominated by hot, blue stars, even though they actually appear red in our telescopes because of the *redshift* from the expansion of the Universe. These are small systems compared to present day galaxies, and relatively simple in form. *Source*: NASA/Hubble Space Telescope.

of the Milky Way. However, it is possible that there is light coming from comparatively older (about 100 million years old) stars that could only be detected with a space-telescope/camera combination that is sensitive to light further into the infrared. If so, these young galaxies are not as small as they appear, and not as simple in structure. At the time of this writing, about one thousand objects like those in Figure 2.11 have been discovered.

The limit of our reach into the *early* Universe is at present about 13.2 billion years, about 500 million years after the Big Bang. Some dozens of galaxies have been located in the deepest images taken with the Hubble. These objects are almost point-like, produced by the most intense places of star formation in the Universe at that very early time. From theoretical modeling, it is expected that the very first stars in the Universe were born around 200–300 million years after the Big Bang, so these detections are remarkably close to what we might call the birth of the modern Universe, as distinct from the earlier Universe, the one composed of only elementary particles and light, a time and state that is not well connected to our origins.

Astronomers find pictures to be powerful first steps to understanding, but most of what we have learned about the cosmos has come from spectroscopy — spreading the light from stars and galaxies, near and distant, in a finely diced rainbow of colors. Each spectrum encodes an enormous amount of information about the physical properties of the source — for example, how hot and dense it is, its *state* (solid, liquid, or gas), its chemical composition — along with dozens of more subtle and related attributes that can be made by combining all the information we gather with our knowledge of physics and chemistry. The Hubble's discovery of nascent galaxies in the first billion years of the Universe is a remarkable accomplishment, but we have learned very little about them beyond their existence. Spectroscopy is crucial if we are to find out their basic properties, but these objects are too faint, for further progress. For the ambitious goal we have set, to follow the birth and evolution of galaxies from the beginning to the present day — a program that has taken almost 100 years of challenging research — even more powerful telescopes are required.

To fill in the pictures of the earliest galaxies with measurements of star formation, chemical abundance, structure and dynamics NASA has built the James Webb Space Telescope (JWST (see Figure 2.12)). The principal goal of this giant, cryogenic telescope is to search for "first light," from the first generation of stars that put an end to the cosmic dark ages. This was an epoch lasting somewhat more than 100 million years in which the fading light of the Big Bang was diluted by the expansion of the Universe. It is a successor to the Hubble that collects almost 10 times the light with its multiple mirrors and is able to see much further into the infrared. The latter requires that the telescope be mind-numbingly cold — 40° above absolute zero, so that the light radiating from the telescope itself does not overwhelm the faint signals from deep space. In equilibrium with the vastness of outer space, any object could cool to such a temperature, but not if it were near the Earth or Moon or illuminated by the Sun, so the Webb telescope will be placed a million miles from Earth

Figure 2.12. The James Webb Space Telescope, the infrared successor to the Hubble Space Telescope. At right, mechanical and optical engineers are assembling the Webb telescope in the giant "clean room" of NASA's Goddard Space Flight Center. The finished mirror of 18 gold-coated mirror segments are mounted on the mirror support structure. The flat pink sheets, approximately the area of a tennis court, are the multi-layer sunscreen needed to keep the telescope mirror in the cold darkness. *Source*: NASA.

and carry a huge "umbrella" that will keep these warm sources safely hidden from the telescope. JWST is one of the most complex machines ever built. Its components are all produced, and assembly and testing procedures are well advanced. The JWST launch is scheduled for 2021.

The cameras on JWST will probe even deeper into the Hubble Deep Fields, with longer wavelengths of infrared light. Its pictures will uncover not just the brightest blue stars in the earliest galaxies, but the (expected) greater amount of mass in Sun-like stars that emit most of their light in the visible spectrum, which for the earliest galaxies is redshifted well into the infrared. For the first time, we will see fully what are the shapes of these first galaxies, how rapidly they are growing, and how much their structural components resemble what we are familiar with in much more mature galaxies. In addition, JWST will have a sensitive near-infrared spectrograph that can record the light, dispersed by color, for a hundred objects simultaneously. This will make it possible to measure, at very early times, accurate SFRs and abundances of the chemical elements, and to search for the evidence of growing supermassive black holes that may actually precede the

The Origin and Evolution of Galaxies

growth of most of the stars. In other words, spectroscopy from space in the near-infrared with JWST will enable astronomers to study the very first galaxies with the same tools now used to study galaxies of today and the recent past. This will complete (we hope) our 100-year-old quest to describe the history of galaxies from birth to the present.

Earthbound telescopes will play a crucial role in the JWST era. One has already been completed — ALMA — the Atacama Large Millimeter/Submillimeter Array (see Figure 2.13), a huge radio telescope composed of 66 individual dish antennae that are spread across the high plain in Northern Chile, over an area reaching 9 miles in diameter. This is what is required to take pictures in radio waves of millimeter wavelength at comparable spatial resolution as the largest optical telescopes and the Hubble Space Telescope. For the study of galaxy evolution, ALMA is opening up a new frontier in mapping in detail the properties of the molecular gas from which stars are born, even for galaxies in the very distant Universe. It is also a powerful probe of the engines of quasars, the monstrous supermassive black holes feeding on gas ripped from stars and galaxies.

What will be the largest optical telescopes ever to be built on Earth are the much needed complements to ALMA and JWST in the study of galaxy evolution. Three huge telescopes, with multi-segmented mirrors amounting

Figure 2.13. (Left) An unprecedented international collaboration, ALMA on the high plains of Chile has begun probing the cooler gas clouds rich in molecules that are the sites of star formation, not just for nearby galaxies, but for galaxies at lookback times of up to 10 billion years. At ALMA's 16,000-foot elevation, above much of Earth's atmosphere, the absorption of millimeter-wave radiation by water vapor in Earth's atmosphere is much reduced. (Right) The Giant Magellan Telescope is also an international collaboration, including eight US institutions plus Australia, Korea, and Brazil. GMT's seven mirrors of 8.4-m diameter will collect enough light to measure the heavy-element abundances and dynamics (stellar motions) of extremely faint galaxies discovered by JWST.

to apertures of 25, 30, and 40 m, (Giant Magellan Telescope, Thirty Meter Telescope and European Extremely Large Telescope) respectively are under construction at this time, all of them international collaborations with billion-dollar budgets. JWST's extraordinary sensitivity to near-to-mid infrared light will enable the detection and study of the youngest galaxies, perhaps all the way back to 200 million years after the Big Bang, when the very first stars are thought to have been born. Combined with the ALMA observations, the star formation histories and the growth of mass and chemical abundances will be obtainable with JWST. However only telescopes much larger than JWST, such as the three mentioned above, can reveal the proportions of different heavy elements — a result of supernovae explosions whose properties could have varied widely, and measure the motions of stars and gas. The much greater light grasp of this next set of telescopes will make possible spectroscopy at higher dispersions (splitting the colors more finely) that is essential to make such measurements. The combination of JWST, ALMA, and these three extremely large earthbound telescopes, has the potential to take us to the final stage of our quest.

The discovery that the Universe is made up of huge, complicated star systems we call galaxies, including our Milky Way, has led to a century of research that has for the first time connected our own evolution and the appearance of life on Earth with cosmic history. Were it not for galaxies containing the heavy elements produced by generations of stars, the complexity required for life — would never have been achieved. It is not hard to imagine a Universe very much like our own in its first 100 million years that *did not form galaxies*, but instead dispersed one or two generations of stars across the vast cosmic sea, and the abundances of carbon, nitrogen, oxygen, silicon, magnesium, iron, etc., too small to lead to planets and life. Galaxies were the reservoirs that held and recycled these heavy elements, allowing their abundances to grow sufficiently to form planets and life. That we are here tells us that our kind of Universe was possible, but by no means *inevitable*.

Our home galaxy, the Milky Way, is reassuringly representative of the most common type of galaxy — a spiral galaxy of moderate stellar mass, large and lacey spiral arms birthing stars, and a small bulge with an impressive but not rapacious black hole. It didn't have to be, since a wide range of galaxy types have abundant stars of the age and mass as our Sun, the long-lived, felicitous host of our planet, so our home galaxy could have easily been one of the much more numerous dwarf galaxies that drift among the giants. Good news for us, because the history of galaxy birth and growth is well exemplified by what happened here, where we can learn about the process in detail, which gives us a leg up on understanding how the variety of galaxy shapes, sizes, and histories — of star formation and chemical enrichment — came to be. Our quest for the full story of galactic evolution from its start to today adds an essential introduction to the story of life on Earth, including our own remarkable chapter.

Further Reading

Christian, C. and Roy, J.-R. 2017 *A Question and Answer Guide to Astronomy*, (Cambridge: Cambridge University Press).

Dunkley, J. 1998. *Our Universe: An Astronomer's Guide*, (Cambridge, Mass.: Harvard University Press).

Elmegreen, D. M. 1998. *Galaxies and Galactic Structure*, (Saddle River, New Jersey: Prentice Hall).

Gallagher, J. S. and Sparke, L. S. 2000. *Galaxies in the Universe: An Introduction*, (Cambridge: Cambridge University Press).

Sheehan, W. and Conselice, C. J. 2015. *Galactic Encounters: Our Majestic and Evolving Star-System, From the Big Bang to Time's End*, Springer-Verlag New York. ISBN 978-0-387-85346-8

Voight, H. H. 1999. *Interstellar Matter, Galaxy*, Universe Portico Content Set: Springer E-Books, Portico Item ID: ark:/27927/pbb7fhktdSpringer-Verlag GmbH E-Book Agreement, Version 1.0 (July 7, 2009) https://access.portico.org/stable?au=pbb7fhktd

William W. H. and Paul H. W. 2003. *Galaxies and the Cosmic Frontier*, (Cambridge, Mass.: Harvard University Press).

Williams, R. 2018. *Hubble Deep Field and the Distant Universe*, IOP Astronomy, Institute of Physics Publishing, ISBN 075031754X, 9780750317542

https://imagine.gsfc.nasa.gov/science/objects/galaxies1.html

https://www.teachastronomy.com/

Chapter 3

The Origin and Evolution of the Chemical Elements

Virginia Trimble

Introduction

A first century sage, asked to explain the law of his people while standing on one foot, is supposed to have replied, "That which is hurtful to you, do not do to another. The rest is commentary." Somewhat similarly, if you ask about the origins of all the elements we know on Earth and in the stars, from gases to rocks and metals, the answer is that hydrogen and helium are left from an early, hot, dense phase in the life of the Universe (called the **Big Bang**); all the rest were made by the stars. Supposedly, the sage then added an instruction to his prospective pupil to go study the commentary. Likewise, this chapter consists of the commentary on the main principle and its extensions to what happens when you look beyond the single stars to the groups of millions to billions to trillions of stars called galaxies, and how those galaxies have changed with time.

Life on Earth is mostly just very complex chemistry. The **chemical reactions** that go on in your body — and in your dog, your potted philodendron, and the mildew in your shower — turn molecules of food, water, and air into flesh and blood, meanwhile producing the energy that keeps you going. To quote Nobel Prize winning physicist Richard Feynman, "yesterday's mashed potatoes are tomorrow's brains," (and this is more obvious in some people than in others).

Origin and Evolution of the Universe

What are the chemical elements? They are the substances that scientists, from the 1700s onward, learned could not be broken down into simpler substances by any process that they could do in a laboratory (though we now know they can be changed by nuclear reactions at very high temperature and density, in the lab, in stars, and elsewhere).

Figure 3.1 shows a periodic table. It shows all the elements we know about: 80 stable ones, found naturally on Earth and in the Sun; a few long-lived unstable ones that linger; and 26 that can now be made only by laboratory reactors and accelerators, and may never have existed naturally. Things you know and love, like gold (Au) and silver (Ar), are stable elements; uranium (U) and thorium (Th) are long-lived but unstable natural elements, and potentially at least very important energy sources; californium (Cf) and bohrium (Bh) are unstable and have lives lasting seconds or less. The last stable element to be discovered was hafnium in 1923. Unstable ones are still being discovered even as we read and write, and now extend the periodic table up to Element 118 (Oganesson, Og).

Not by chance, the chemical elements most important in living creatures are, with a few exceptions, the commonest ones in the Universe, beginning with hydrogen and oxygen. Their compound, water, makes up half or more of the weight of most cells. Next is carbon, whose unique ability to combine in many different ways with many other atoms makes it the building block of most organic molecules, including carbohydrates and fats. Nitrogen is an essential part of proteins. Other common, important elements are the iron in your blood and the calcium and phosphorus in your bones.

Some less common elements, like manganese, selenium, magnesium, sodium, chlorine, and potassium, are also essential for long-term health and are found in a variety of plant and animal tissues. An easy way to identity out essential trace elements is to look at the label on the bottle of a good brand of multivitamins and minerals. The heaviest essential element appears to be iodine for thyroxin. Sulfur contributes to the coloration of some plants and bacteria (not all of which cause diseases!). The only common chemical elements not found in living cells are helium, neon, and argon, which form no useful compounds, and so are called *noble gases*.

Clearly, life would have developed very differently (if at all), if some different set of elements were the common ones on Earth and in the Universe. Thus, it becomes interesting to ask why there is lots of oxygen and carbon, but very little beryllium and fluorine. Along the way, we will also find out where lead, arsenic, mercury, and other elements that are

The Origin and Evolution of the Chemical Elements

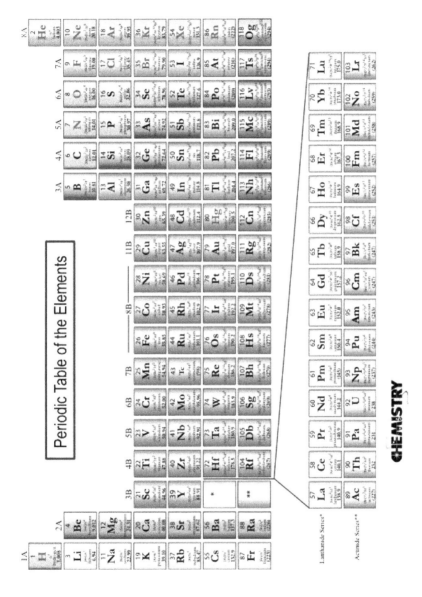

Figure 3.1. Current version of the Periodic Table of the Elements. *Source:* American Chemical Society.

definitely not good for you come from. It should be said that we now understand element building (the fancy name is **nucleosynthesis**) well enough to be able to say that a Universe that made more fluorine than oxygen (though they are neighbors on the periodic table) or more arsenic than phosphorus (neighbors in the other direction) would be different from ours in many other ways that might well keep stars from shining as long as life needs them to, or otherwise be uninhabitable. Some astronomers and physicists actually worry about this sort of thing, and call the subject "fine-tuning" or "the anthropic principle." These will not be examined in this chapter.

In fact, in our Universe, we understand reasonably well what nuclear reactions can happen where and when, and what they produce, so that stellar structure and evolution and nucleosynthesis are among the better-understood parts of modern astrophysics. They are also among the longer-established parts, so that much of what is said here could have been said (indeed was said) several decades ago.

The following sections begin with what was said decades ago, broad-brush changes and improvements, and then go on to various astronomical sites of nucleosynthesis and what is produced in each, ending with an overview of the evolution of the chemical composition of our galaxy and its implications for numbers, ages, and locations of **habitable planets** from among the thousands now known to orbit other stars. Figure 3.2 shows the measured abundances of the elements in the solar system as a function of the number of protons in the nuclei of their atoms. If we have understood the origin and evolution of the chemical elements, then all the processes will add up to this average. By and large they do.

By way of reminder, chemical reactions merely move atoms of particular elements in and out of molecules of particular compounds (for instance, burning coal or carbon in air or oxygen to make carbon dioxide). Nuclear reactions, in contrast, turn one element into another, as when three helium atoms fuse to make a carbon atom. If you care to look inside the atoms (see Figures 3.3–3.5) you will see that chemical reactions involve borrowing, lending, and sharing of electrons in the outer parts of atoms, while nuclear reactions attack the protons and neutrons inside, sometimes even turning protons into neutrons or vice versa.

One final refinement is needed at this stage: many elements (not all) come in variations, such that every atom has the same number of protons but not always the same number of neutrons. These are called **isotopes**,

The Origin and Evolution of the Chemical Elements

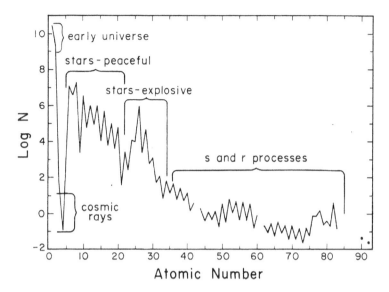

Figure 3.2. Relative abundances of the naturally occurring long-lived and stable elements, from hydrogen at the left to uranium at the right. The vertical axis is logarithmic, so that hydrogen is a million-million times as common as beryllium (atomic number = 4). The types and locations of the nuclear reactions that produced each group of elements are shown.

and can be stable or unstable. The total number of protons plus neutrons is called the atomic weight, and Figure 3.2 shows abundances of the stable isotopes of elements known to occur naturally on Earth.

Why should "we" care? As biologists we do not care so much. Though living creatures are used to the mix of isotopes we grew up with, generally it shouldn't matter, because chemical and biochemical reactions are determined by the electrons, hence the protons. This it must be said, has not really been tested as far as I know; but with one exception. Hydrogen with two neutrons (called **deuterium**) has twice the weight of ordinary hydrogen, and though fish are fine with the standard Earth mix (one deuterium atom for every 10,000 or more hydrogens), they are said to die in bowls filled only with heavy water (or deuterium oxide).

As geologists and archeologists, we might care, because some unstable isotopes (of uranium, thorium, potassium, and, best known, carbon — called carbon-14) can be used to date rocks and samples of wood and cloth — date,

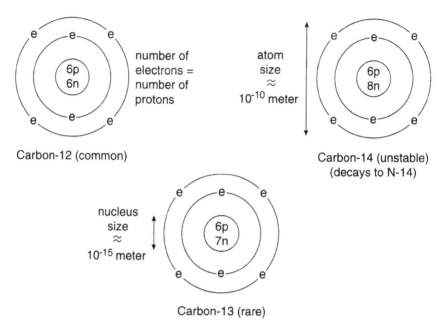

Figure 3.3. Structure of elements, atoms, and isotopes. Each element consists of atoms having a particular number of protons in their nuclei. Carbon has six. Isotopes are variants of elements with different numbers of neutrons in the nuclei of their atoms. Common carbon has six neutrons; a rarer kind has seven; and you can tell organic from inorganic carbon in old rocks by the amount of each. An unstable isotope, carbon-14, decays to nitrogen-14 in about 5700 years. It is made by cosmic rays hitting the Earth's atmosphere and is used to date archaeological sites.

in this context, meaning measure the length of time since a rock became solid or since a tree was alive.

As astronomers, of course, we care, because we are perfectionists. We want the nucleosynthetic processes that follow to add up, not just to the mix of elements we see but the mix of isotopes. Obviously, it all works out, or I wouldn't be telling you about it! But it isn't the simplest possible story, because many elements have their isotopes made by different processes. For instance, among the ones to follow, ordinary carbon, C-12, comes from three helium atoms fusing in evolved stars; C-13 (about 1/90th of Earth's carbon) from a kind of hydrogen burning in stars more massive than the Sun; and C-14 (the unstable kind used for dating) is made in earth's upper atmosphere by **cosmic rays** hitting nitrogen. It decays, incidentally, back to ordinary N-14.

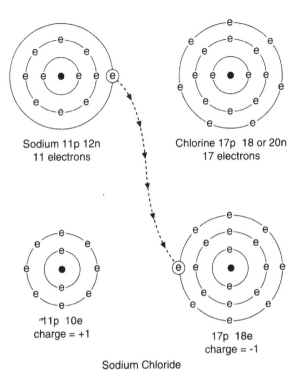

Figure 3.4. Chemical reactions, including biochemical ones, are reactions in which only the electrons are involved. They can be lent, borrowed, or shared, until the outer shell contains a stable number of electrons (2, 8, 8, or 18, depending on the element). The atoms involved are then held together by the unbalanced charges. In this example, an atom of chlorine takes an electron from an atom of sodium, leaving each with a closed outer shell of eight electrons. The plus (+) charge on the sodium and the minus (–) charge on the chlorine hold them together in a molecule of sodium chloride (ordinary salt).

Historical Overview

What we think we know today about the origin of the elements rests on two pillars. First, Burbidge *et al.* (1957) wrote a 100-page review article called *"The Synthesis of the Elements in Stars."* It included what was known at the time about how much of each element was present in our solar system and a discussion of the nuclear reactions that could occur in stars to produce those elements. They chose not to include reactions in the early, hot, dense Universe (Big Bang) because they were, to varying degrees, supporters of a steady-state model of the Universe, in which

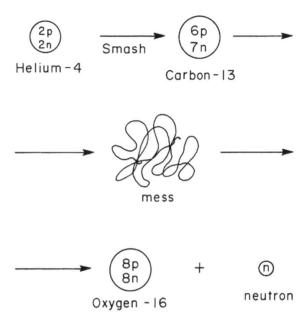

Figure 3.5. Nuclear reactions are ones in which protons and neutrons move between atomic nuclei or even change into each other. In this example, a helium-4 strikes a carbon-13; the particles are thoroughly entangled; and there emerge an oxygen-16 and a single neutron. The neutron is unstable and will decay to a proton + electron + neutrino unless it is captured by another nucleus within about 11 minutes. This reaction is the source of neutrons for a subset of the capture by iron (etc.) that build up the heaviest elements.

there never was an early, hot, dense phase. Apart from that, their assortment of processes and sites for those processes comes very close to our best modern understanding of what has gone on, and a 50th anniversary celebration at the California Institute of Technology in 2007 was enormous fun, although only two of the four (the Burbidges) were alive to participate (see Figure 3.6).

Very crudely, down to the present time, we have concluded that the early universe was important, that the set of seven processes identified by B²FH can vary a good deal from one sort of star or stellar explosion to another, and that there are mixes of elements in other stars (especially old ones) and galaxies that are fairly different from the mix in our Sun and similar stars. The trick now is to tune the assortment of processes to match the assortment of compositions.

Second, the 1968 Ph.D. thesis of Beatrice Muriel Hill Tinsley forced upon us the realization that the evolution of whole galaxies could be

Figure 3.6. Pioneers of nucleosynthesis theory. Left: the authors of B2FH: Margaret Burbidge, Geoffrey Burbidge, William Fowler, and Fred Hoyle. Right: Beatrice Tinsley.

calculated and that the changes were important. Her assumptions were rigorous but not unreasonable: (a) she divided up stars into mass groups (0.5–0.85 to bigger than 10 solar masses) and said that all stars in each group would do the same things; (b) star formation was approximated by bursts each of a billion years, and because most of the heavy elements come from massive stars, the heavies were assumed to be added to the model galaxy as soon as the stars formed; (c) each burst included a certain percentage of stars in each mass group, and that percent was the same for all bursts; (d) the new heavy elements were assumed to get mixed in uniformly through all the remaining gas; and (e) no material was allowed to flow in or leak out.

In summary, this was a "homogeneous, constant **initial mass function (IMF)** [distribution of number of stars by mass at birth], instantaneous recycling, closed box model. It told us two things. First, this relatively simple calculation (though it strained the computers of the time) would provide quite a good match to the observed properties of galaxies of different brightnesses, colors, compositions, and residual gas fraction. Second, evolution mattered. Galaxies would generally have been brighter and bluer in the past, when they had more young stars, and so could not be used in the simplest possible way to determine the expansion rate of the Universe and whether that expansion was speeding up or slowing down. The next round of models allowed for variable IMF, noninstantaneous recycling, incomplete mixing, and leaky boxes, most in aid of what was long-called the G-dwarf problem, which will be explained to you in due course.

Even more crudely, what we have learned since is that the galaxies we see here (meaning within the nearest billion light years or so) and now (meaning that we see light that left them not much more than a billion years ago) did not start life as Beatrice's giant clouds of gas. Instead, the first things that formed were small blobs of **dark matter**, not made of any of the elements we know about (see Chapters 1 and 2). These contracted with masses perhaps only a million times that of the Sun. Hydrogen and helium gas flowed in and formed the very first stars, which began dying and scattering heavy elements around. Meanwhile, the small dark matter blobs (carrying stars and gas with them) gradually got together (gravity always wins!) to make bigger blobs like our Milky Way. The oldest star clusters in the Galaxy (called "globular" for their uninspiring shapes, see Figure 3.7) probably formed in the long-ago dwarf galaxies from which the Milky Way was later assembled.

There are still some of those small galaxies around to study, and the Milky Way and other galaxies are still growing by sweeping them up to add to their inventories of stars, gas, and dark matter. Curiously, dark matter hardly enters the picture at all, except to hold gas and stars together. It is not capable of nuclear reactions, or even shining or blocking our view of things. The dark patches you might see against the light of the Milky Way (if you live somewhere with dark skies) are just clouds of gas (essentially transparent) containing 1–2% dust (quite opaque — think smoggy days or a dust storm in the desert).

To see how this might work out, let us do a small calculation. A proper calculation of the building up of the elements (nucleosynthesis) and galactic chemical evolution requires everything we know about (and some things we don't) about a very wide range of subjects. These include (a) star formation, (b) structure, including **convection** (orderly gas flows that transport energy and mix gases of different composition and temperature) and turbulence, (c) the **opacity** of gas, (d) how nuclear reactions depend on gas density, temperature, and composition, (e) how stars sometimes blow off their outer layers peacefully, but sometimes also explode as novae and, especially, **supernovae** (see Figure 3.8), distributing the heavy elements they have made, (f) acceleration of high-energy particles called cosmic rays, and more.

But consider our galaxy. In addition to the dominant, almost 10^{12} solar masses of dark matter, it contains a bit less than 10^{11} solar masses

Figure 3.7. Hubble Space Telescope image of the globular cluster in Scorpius Messier 80 (NASA). This cluster is 30,000 light years away and contains several hundred thousand stars. The field of this picture is about a hundred light years across. The typical separation of adjacent stars inside this cluster is still very large (a tenth of a light year) — their images only appear to be touching because of the limited spatial resolution of the Hubble Space Telescope. Note that most of the brightest stars are the red giant stars, in their final stages of thermonuclear burning.

of stars and gas. Somewhat less than 1% of this consists of heavy elements, so we need to have accumulated something like five hundred million solar masses of heavies over the age of the Universe. A typical supernova gives you a few solar masses of stuff beyond H and He, so we need perhaps two or three hundred million supernovae in total, independent of whether these occurred in the fully assembled galaxy, recently, or in the component parts early on. This comes to a couple of SNe per century, which is just about what is observed in galaxies very much like ours.

Figure 3.8. False color image of the supernova remnant Cassiopeia A (also shown on the cover of this book). Red shows infrared radiation from heated dust grains. Orange shows visible light. The green and blue colors show X-rays from hot and super-hot gas. The cyan dot in the middle is the remnant of the star's core. *Source*: NASA.

We don't have a very good view of our own inventory, because we are down in the dusty murk of the galactic disk, (fish don't make the best ichthyologists either), but the last few supernovae for which we have good evidence were in 1006, 1054, 1572, 1604, and about 1680, mostly from Chinese astronomers. The light, of course, left those events thousands of years before those dates, but in any case, it would seem that we might be a bit overdue for a galactic supernova, which, if nearby, will be truly spectacular. The closest stars to which this can happen in the next 10,000 years or so are Antares and Betelgeuse. Both are much too distant for their flash of ultraviolet (UV) light and X-rays to be dangerous, but brightness like a full moon is plausible.

At this point, we have some grasp of the sages' claim: hydrogen and helium come from the Big Bang; Burbidge, Burbidge, Fowler, and Hoyle (or stars if you prefer) made all the rest. We can go on to look at some of the details, both the well-established and the puzzling.

Big Bang Nucleosynthesis

The 19th century chemists who clarified the numbers and identities of the elements and started to figure out how much there was of each (at least on Earth and in the meteorites) were also the first modern speculators on our origins. The majority view was (Trimble, 2010) that the simplest atoms (hydrogen) came first, with a subset settling down into heavier elements in some widespread process before any of the stars and planets we now see had formed.

The 20th century revolutions of quantum mechanics and general relativity swept across this peaceful picture, leaving only two sites hot and dense enough for nuclear fusion. First was the whole Universe close to its origins, in which case cosmic composition was fixed before the stars came. The other was the centers of stars both now and in the past, in which case hydrogen fusion would be an ongoing process, perhaps observable. Both were indeed considered, but the devastating facts of two world wars and a nearly worldwide depression in between heavily affected who could think about what, when.

One person actually rode both horses, the Russian–American physicist George Gamow. He moved from his hometown of Odessa to St. Petersburg (Petrograd/Leningrad) to work with the cosmologist A. A. Friedmann, who then died almost immediately (not causal). Gamow then moved on to Germany and an attempt to unify Einstein's general relativity theory of gravity with electromagnetism, which cannot be done. Embracing quantum mechanics, he wrote a doctoral dissertation that included a mechanism that would allow atomic nuclei to break up into other elements, even when they didn't have enough energy to do this according to classical mechanics. In accordance with the general principle that most things are easier to get into than out of, the same mechanism allows fusion of smaller nuclei into bigger ones with less than the classically necessary energy. It is called barrier penetration or tunneling, and is an essential part of how we understand nucleosynthesis today.

The pioneering discussions of possible conversion of hydrogen to helium came from people Gamow left behind when he moved again, onward to fellowships in Copenhagen and Cambridge. That track leads to stars (see the following next section), while George moved still further west to the United States and, starting in 1935, considered the possibility of building up heavy elements in an early hot Universe (later called the Big Bang).

Gamow's (1949) idea survived World War II, which he spent as a professor at George Washington University. Soon after, he teamed up with younger physicists, Ralph Alpher and Robert Herman, to follow the evolution of a Universe that began as pure neutrons. But, this was wrong. The starting point has to be a thermal equilibrium of protons, neutrons, electrons, light, and neutrinos. Curiously, they got just about the right answer for the emergent helium fraction. I have always regarded this as something of a mystery! They expected to be able to build up all the elements from repeated neutron captures, and predicted that the result would look a good deal like Figure 3.2.

But there was a fatal flaw! No stable atom has either five or eight protons plus neutrons. So, they could make ordinary hydrogen, helium, plus small amounts of lithium, deuterium, and helium with only one neutron, but there was a great chasm between them and anything heavier. But in the process of considering a neutrons-first Universe, Alpher and Herman predicted that cosmic nucleosynthesis would leave behind large quantities of X-rays, which would expand and cool with the Universe, until the present, when the radiation would have a temperature, they calculated, near 5°K. That it might be observable did not occur to them at the time.

The prediction was temporarily lost; Gamow went on to try to crack the genetic code. The radiation was discovered largely by accident in 1965 (Chapter 1). The associated Nobel Prize went to the discoverers, in 1978. Gamow (who lived until 1968) was at a 1967 conference where the importance of the radiation was extensively discussed. I was there, too, and share with other participants the feeling that he was not entirely persuaded of what had been found, on the grounds, said he "I lost a nickel; you found one (remember that 5°K). Who's to say it's the same nickel?" Alpher and Herman, who lived considerably longer, expressed some displeasure in not sharing the Nobel Prize.

Not surprisingly, the exact amounts of ordinary hydrogen, deuterium, ordinary helium, and helium with only one neutron (He-3) you get depend very sensitively on the exact temperature–density history of the early Universe. Are scientists today frantically trying to improve the gory details of Figure 3.9? Not really. The whole picture of that cauldron and its relics has today been subsumed in what is called "concordance **cosmology**" (Chapter 1). So, although Big Bang nucleosynthesis was the first consideration to tell us firmly that most of the dark matter (Chapters 1

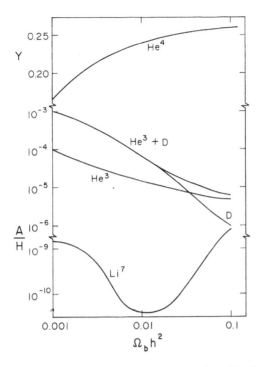

Figure 3.9. Abundances of the elements and isotopes produced in the hot, dense early Universe (Big Bang) as a function of the density of ordinary matter participating in the reactions. On the horizontal axis, 0.01 means 1% of the density needed to stop the expansion of the Universe, for a Hubble constant of 100 km/s/Mpc (h = 1). Helium-4 is shown as a fraction, Y, of the total mass of ordinary matter (a little less than 25%), and the others as ratios of numbers of their atoms (A) to numbers of ordinary hydrogen (H), for deuterium (hydrogen-2, also called D), helium-3, and lithium-7. The lithium curve has its odd shape because the element is made by two different reactions under different conditions. Observed values of the abundances of ^{2}H, ^{3}He, ^{4}He, and ^{7}Li in material that has not been exposed to nuclear reactions in stars show that the density of ordinary matter (now hydrogen, helium, carbon, and all the rest) is much less than would be needed to stop the expansion of the Universe in the distant future. *Source*: The Astrophysical Journal.

and 2) could not be made of anything consisting of protons and neutrons, and that certain kinds of hypothetical exotic particles didn't lurk back then either, the same information is now available other ways.

From the point of view of a habitable Universe, any Big Bang that left enough hydrogen for stars to live on would do. Oddly, the scrap of lithium (one atom in ten billion) that is also formed is important. To turn gas into stars, you must cool it down to temperatures well below freezing.

This is easy to do with gas today, with its 1–2% heavy elements. Those, especially carbon and the molecules they form, are good at radiating away heat as visible light, infrared and microwave radiation, thus cooling the gas. Pure H + He does much less well, and a few atoms to make lithium hydride (LiH) help a lot (Galli and Palla, 2013).

Element Synthesis in Stars

To progress beyond hydrogen and helium, we must find some other location that is about as hot and dense as the early Universe. One such place is a nuclear reactor or accelerator in a physics laboratory. And, indeed, virtually all of the reactions discussed here can be duplicated in laboratory experiments. The results of the experiments are measurements of how fast particular reactions go at different temperatures and densities, the amount of energy released (or absorbed) in the reactions, and the particular elements made by each. These results fit nicely into a theoretical framework that comes from a general understanding of the forces that hold protons and neutrons together in nuclei. And the product elements are the ones we see in stars. This sort of interplay among experimental results, theory, and observations is characteristic of the way astronomy and other sciences progress. Because things fit together, we can be confident we understand which nuclear reactions will occur under any set of conditions and what they will produce. Another place that is hot and dense enough for nuclear reactions is the center of a star. That stars derive most of their energy for most of their lives from nuclear reactions has been known for about 90 years, and Hans Bethe (Nobel Prize in physics, 1967) wrote down the details of the most important ones in the late 1930s (Bethe and Critchfield, 1938; Bethe, 1939).

Understanding of the more complicated reactions that make the heaviest elements came more slowly after World War II. A key event was the 1957 publication, by Burbidge *et al.* (1957), of a massive review of observed element abundances in stars and of the full range of nuclear reactions needed to make them all. This paper is important enough in astronomers' minds that it is instantly recognized by the acronym B^{2}FH, mentioned earlier. Since then, studies of nucleosynthesis have fine-tuned their results, and more recent reviews of the subject (Trimble, 1975, 1991) normally build upon the framework erected by B^{2}FH. Even now, more precise numbers are still needed for a few reactions (for instance carbon + helium making oxygen), and these are being actively pursued by experimenters and theorists in many countries.

Figure 3.10. Stars can live for billions of years on hydrogen fusion precisely because the Big Bang made 75% or more of the matter in the form of hydrogen. That ordinary stars are made mostly of hydrogen and helium was first shown in the Ph. D. dissertation of the woman shown here, Cecilia Payne, later Payne-Gaposchkin. It is exceedingly rare for a scientist to make a fundamental discovery as part of a thesis project (though you will hear about one other case later), and Payne's results were not fully believed by the community until the dominance of hydrogen had been demonstrated by another, more senior astronomer. The picture, which resembles a Coptic tomb portrait, was drawn by her husband, Sergei Gaposchkin, in about 1945 (image courtesy of Katherine Gaposchkin Haramundams).

The Major Burning Phases — Hydrogen

Let us start with an outline (see Table 3.1). It will be easy to remember if you keep in mind that the lightest nuclei, with the fewest particles in them, generally fuse at the lowest temperatures (beyond iron you cannot fuse them at all), and stars evolve with time in the direction of having hotter, denser cores.

The first set of reactions, and the ones that keep stars shining for 90% or more of their lives, simply fuse hydrogen into additional helium. Next, helium burns (a standard synonym for **fusion** reactions) into carbon and oxygen. Our Sun will progress no further. In more massive stars, carbon- and oxygen-burning makes neon, magnesium, silicon, and other intermediate-mass elements, which, in turn, fuse to iron and its neighbors (nickel,

Table 3.1. Stellar evolutionary stages.

Reactions	Products	Temperature of burning (K)	Time scale for Sun (yr)	Time scale for star of 20 times mass of Sun (yr)
Hydrogen-burning	Helium	$1-4 \times 10^7$	10^{10}	10^7
Helium-burning	C, O	$1-2 \times 10^8$	10^9	10^6
Carbon-burning	Ne, Na, Mg	8×10^8	—	300
Neon-burning	Mg, Si	1.7×10^9	—	< 1
Oxygen-burning	Si, S	2.1×10^9	—	< 1
Silicon-burning	Ti to Zn	4×10^9	—	2 days

copper, and others). At this point, the star is in serious trouble. So far, each reaction stage has released energy to keep the star hot and shining. The protons and neutrons in iron nuclei are grouped as tightly together as the forces permit, and no further nuclear energy can be extracted. The outcome is one kind of supernova (see Chaper 4). And the building of elements beyond iron occurs in a rather different way.

Hydrogen burns in two separate reaction sequences, both known to Bethe. In the simpler process, two protons come together, one turns into a neutron, and they stick together to give a deuterium. The deuterium nucleus quickly picks up another proton, and, after a few more particle collisions, you have a helium-4 nucleus, and lots of energy, which the star then radiates. This **proton-proton (or p–p) chain** has been the main energy source in our Sun for the past 4.5 billion years, and will continue to be so for the next 5 billion or thereabout. Two interesting sidelights: first, the p–p chain begins by turning a proton into a neutron. This is a very slow reaction, and the Sun therefore burns its hydrogen quiescently. In hydrogen bombs, you start with deuterium, and the rest goes very quickly. Second, turning protons into neutrons releases tiny, uncharged particles called **neutrinos**. These stream directly out from the center of the Sun (while the electromagnetic energy takes 100,000 years to get out) and are detected here on Earth. Thus, we know that the Sun is fusing hydrogen right now.

Hydrogen can also burn in what is called the **CNO tri-cycle**, in which atoms of carbon, nitrogen, and oxygen act as catalysts. Of course this can happen only if C, N, and O atoms are already present, which means that the very first generation of stars had to begin with the proton-proton chain in all cases. But stars today have about 1% of their masses in CNO, and the CNO cycle is the main hydrogen burner for stars more than about twice the mass of our Sun. It is interesting because a by-product is the conversion of some carbon and oxygen into nitrogen, which otherwise does not turn up in our outline, and is vital for **proteins** and **DNA**.

Hydrogen-burning continues until the central 10% of the star is all helium. This takes billions of years for smallish stars like the Sun, but only millions of years for more massive ones. At the same time, nuclear reactions are changing the composition of a star's center, its outside reacts by expanding or contracting and changing temperature. We can calculate what those changes will be, and compare them with the brightnesses, sizes, and colors (temperatures) of stars we see at different ages and masses, thereby checking that we have the right answer. Very broadly, an evolving star gets brighter and redder for most of its life, with the possibility of one or more switchbacks to being fainter and/or bluer. Hydrogen-burning stars are said to be on the **main sequence**, and more evolved ones are called **red giants** and **supergiants**.

The Major Burning Phases — Helium

So far, we have made helium. This was the point at which the Big Bang hung up. What is the difference? The Universe cools as it expands, and particles get further and further apart, so it is harder and harder for them to interact. Stellar cores, in contrast, get hotter and denser. And by the time 10% of the interior has burned to helium, it becomes possible for three helium nuclei all to meet up within a tiny fraction of a second and fuse to form carbon.

Helium-burning starts explosively in low-mass stars and shakes them up, but starts peacefully in massive stars. Similar cases occur later, so we might as well bite the bullet and understand why right now. Ordinary hot gases try to push outward and expand; and the hotter they are, the harder they push. Normal nuclear reactions release energy and heat the burning gas. If a reaction goes too fast, the gas overheats and expands, cooling back down again to stability. But in very dense gases, the amount of pushing does not depend on temperature. Such gases are said to be **degenerate** (a statement about how their electrons are moving, not about their moral principles). If you turn on a nuclear reaction in a degenerate gas, the gas heats up, but does not expand. The hotter gas burns faster, and so forth, and pretty soon you have a nasty explosion on your hands (or at least in your star). It is the low-mass stars that start helium-burning with a flash, because they are denser than big stars. This probably sounds backward, but it really is the right answer.

Helium-burning has two parts that go on together. Three He nuclei fuse to a single ^{12}C; and ^{12}C plus ^{4}He makes oxygen-16. The rate of this second part is very important (and not very well known), because the relative amounts of C and O made determine what is available to burn later. That you get some of each is vital to our existence on Earth — we need both carbon-based food and oxygen-rich air!

At the same time helium-burning is making C and O for us in stellar cores, hydrogen fusion continues further out. The stars are by now very bright and are blowing large amounts of gas off their surfaces in winds we can see (even the Sun has a very weak one). Both nuclear reactions and the vigorous loss of mass go on until one of two things happens. Either the C–O core gets hot enough for further reactions (of which more in a moment), or wind loss strips the star down to its core, which then begins to cool, turning off hydrogen- and helium-burning. Additional reactions are the fate of stars that begin life with more than about 8 times the mass of our Sun, and stripping ends the lives of the others.

Figure 4.1 (Chapter 4) shows a product of the stripping operation. The hot, remnant central core illuminates the surrounding material blown out in the wind. The gas cloud is called a **"planetary nebula"** for misleading historical reasons (nothing to do with whether the star had planets). The core will cool to what is called a **white dwarf**, and, in general, do nothing else interesting for the rest of the history of the Universe.

From a nucleosynthetic viewpoint, most low-mass stars are rather a disappointment. They keep most of the C and O they have made inside the white dwarf. The wind does, however, carry off some carbon and nitrogen and perhaps oxygen, and products of one minor set of reactions (see below) and dumps them back into the general **interstellar medium**, where new stars are forming (see Chapter 5).

The winds are most vigorous, and give us the most carbon (etc.) very late in the stars' lives, when they are fusing both hydrogen to helium and helium to carbon and oxygen in thin shells around C–O cores (Figure 3.11). The obvious name for this is double-shell-burning! Less obvious, but more commonly used by astronomers is "asymptotic giant branch phase."

Very rarely, a white dwarf (or a pair of them) left behind by dying, low-mass stars gets into trouble. No white dwarf can exceed a mass 1.4 times that of our Sun, or it will either collapse or start nuclear burning and explode (because it is degenerate; I told you we would need this again!). A previously stable white dwarf can find itself above the limit if another star dumps material onto it or if a close pair of white dwarfs merge. In

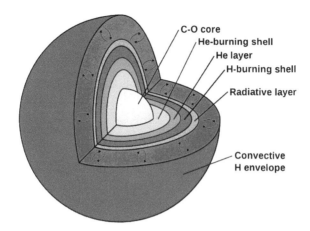

Figure 3.11. Cut-away view of solar-mass star at the end of its lifetime, before producing a planetary nebula (by ejecting the outer layers) and white dwarf remnant (from the bare C–O core). *Source*: National Optical Astronomical Observatory.

either case, the result will be a kind of supernova (see Chapter 4). The explosive burning turns most of the white dwarf into iron and related elements. Such events are the main source of iron (plus chromium, cobalt, etc.) at some times and places in the Universe. These are also the supernovae used as measuring sticks for distant parts of the Universe, revealing that cosmic expansion is now speeding up (as discussed further in Chapter 4).

The Major Burning Phases — Heavy Elements

In very massive stars, wind stripping still occurs. But it does not prevent the onset of carbon-burning. From now on, the star uses up its fuels so quickly that the outer layers never find out what is going on inside. The star continues to look like a red or blue supergiant, whichever it was when carbon fusion started. At successively higher temperatures and with successively shorter time scales, carbon-burning is followed by neon-, oxygen-, and silicon-burning. The durations are brief, partly because these reactions release rather little energy compared to hydrogen- and helium-burning, and partly because the very hot stellar cores radiate enormous numbers of neutrinos, which drain away energy without contributing to starlight.

Another factor common to the reactions that burn heavy elements is that all are complex networks, rather than simple chains or cycles. Several

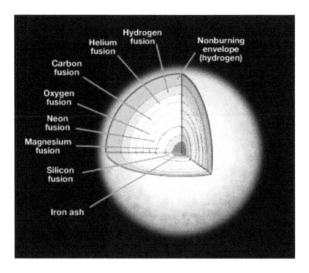

Figure 3.12. Artist's illustration of the core of a massive star just prior to a type II supernova explosion. The core is a series of nested spherical shells, with each shell fusing a different element from hydrogen to helium, to carbon, through the periodic table to iron. Note that in between the shells that are actively fusing elements, there are inert regions in which fusion is not occurring. Reprinted from Penn State Astronomy & Astrophysics.

different elements, as well as free protons, neutrons, and helium nuclei, are all present at once and can interact in many different ways. Each burning stage thus makes many different elements and isotopes, of which only the commonest are mentioned in Table 3.1. As each new reaction turns on at the center of the star, the others move outward into slightly cooler regions and continue to operate, until the star looks rather like an onion (see Figure 3.12).

The onion skins work their way outward through the star until the mass of the central iron (etc.) core grows to a little more than the mass of our Sun (remember this is a star that began with 10 or 20 or 30 solar masses). Catastrophe is at hand. There is a maximum stable mass for the degenerate iron core, just as there is for degenerate white dwarfs. The limit is called the **Chandrasekhar mass**, curiously enough because it was discovered by Chandrasekhar (1931, 1935) (You can't count on these things — Hubble's Law, discussed in Chapter 1, was discovered by K. Lundmark.) In any case, Chandrasekhar shares some tiny part of the blame for this Chapter, because he was the teacher of the present author's thesis advisor. Chandra's Nobel Prize arrived 53 years after his landmark paper was published.

A vital difference between massive stars and the ones that make white dwarfs is that the white dwarf keeps most of the carbon and oxygen made by its parent star, while supernova explosions of more massive stars blow off most of the heavy elements they have made. The precise amounts of each element and isotope in the onion, waiting to be blown out, depend on the mass of the star and other factors. In addition, those amounts will change after the shock of the explosion passes through the layers, generally making a bit more of the less common isotopes. But the general patterns are clear and agree with what we see as the composition of the Sun and stars.

On average, you get the most of the things that form first. Next, elements with even numbers of protons are commoner than ones with odd numbers. Neon outnumbers fluorine and sodium; silicon outnumbers aluminum and phosphorus. Isotopes with even numbers of both protons and neutrons (ones you could think of as being put together from a bunch of helium nuclei) are particularly abundant — ^{16}O, ^{20}Ne, ^{24}Mg, and ^{28}Si. And there is lots of ^{56}Fe. These patterns make sense. The commonest species are the ones that, according to our understanding of nuclear forces, have their protons and neutrons held together most tightly. Similar patterns will appear among the heavier elements discussed in the following section.

Massive stars are much more important for nucleosynthesis than smaller ones like the Sun for three reasons. First, they experience a much wider variety of nuclear reactions with a much wider range of products. Second, at the ends of their lives they blow out most of the heavy elements they have made in supernova explosions, whereas stars of 8 solar-masses or less keep most of their products in the C–O white dwarf remnant. Third, their lives are much shorter. Thus, many generations of 30-solar-mass stars lived and died before our solar system ever formed, and many more have formed and exploded since.

On Beyond Iron

In a sense, we are almost through. Elements whose origins you have now heard about make up 99.99997% of all the matter in the Sun and other stars. On the other hand, we have made only the first 30 elements, and there are another 62 to go, before we get to uranium. Admittedly, not all are stable, and so not all are currently found in the Sun or Earth or meteorites. On beyond uranium are 18 or more short-lived elements that have been produced in terrestrial laboratories. Of these, at least plutonium was present when the Earth first formed and so must also have been made in nature.

Where did these few, but highly varied, atoms come from? Ordinary fusion reactions are no longer relevant. Throw two iron nuclei (each with 26 protons) at each other, and they will either bounce off or break each other up, not fuse to make tellurium, the element with 52 protons. As B²FH showed, three additional processes are needed to make the full range of elements beyond number 30. In two of them, seed nuclei, mostly iron, capture neutrons, painfully, one by one, with neutrons occasionally decaying to protons, to keep the nuclei stable. A third process acts on the products of these neutron captures (from germanium to lead) and either adds protons or knocks away a few neutrons. No wonder these products are rare! Rarest of all are those from the third process, ranging down to tantalum-180, of which there is one atom for every 10^{16} atoms of hydrogen in the Universe!

Recognizing the need for two separate neutron capture processes was a real triumph, also achieved by Cameron (1957), though his version was not publicly available until later than B²FH. Remember Gamow, who wanted to make everything out of neutrons? His evidence for this was the correlation between abundances of elements and isotopes and their willingness to capture another neutron. Unwilling capturers pile up and become abundant, while willing ones are quickly turned into other species and remain rare.

That piling up around the most tightly bound nuclei is the same for the captures we now know happen in stars as it would have been for Gamow's pure-neutron Universe. That's why he was able to persuade himself for many years that he had the right picture of things.

But careful examination of abundances of Elements 30 to 92 reveals two patterns. Some relatively common isotopes are unwilling to capture neutrons in their present condition. They have 50, 82, or 126 neutrons now ("magic" numbers because they are closed neutron shells, analogous to shells of 8 or 18 electrons). Barium-138 and lead-208 are examples. Other relatively common isotopes would happily capture neutrons now, if they had a chance. But suppose they formed by grabbing every neutron in sight, while they could, long ago, until they had 50 or 82 or 126, and then some of those neutrons later decayed to protons, leaving atoms of greater long-term stability. This would account for the relatively large amounts of tin and tellurium, rhenium, osmium, and platinum now found. I keep emphasizing the "relatively" because even the commonest of these elements makes up less than one atom in a billion.

The Origin and Evolution of the Chemical Elements

B^{2}FH called the two necessary processes s (for slow neutron capture) and r (for rapid neutron capture), where slow and rapid mean in comparison to the time it takes an unstable nucleus to decay back to a stable one, by having a neutron turn into a proton. This time ranges from minutes to years. The third, fine-tuning process is generally called p, for proton. How and when and where do these three processes happen?

One site of the r- and p-processes is the various kinds of supernova explosions, when indeed both lots of iron and lots of neutrons will be available together, at least for a short time. Just how you get the products out of the explosions without damaging them is a topic of current debate. Uranium and thorium, without which atomic bombs could never have been developed, are among the elements made only by the r-process (oh, sorry; I said I wouldn't mention bombs again!).

The other site is, I think, more fun. Some neutron stars occur in close pairs with other neutron stars or perhaps black holes. Neutron star binaries have been already been known for several decades (see Chapter 4). The existence of black hole binaries was recently proven by the remarkable detection of gravitational waves from the last second of their merger process. This extraordinary discovery of the gravitational signal from the mergers was only possible by great expenditure of manpower and resources, in America, Europe and other countries, in constructing the **Laser Interferometer Gravitational-wave Observatory (LIGO)** (see Chapter 4). The pairs gradually spiral together (again this is seen), and the stars eventually merge, making a nasty mess, difficult to calculate! Part of the stuff from the binary black hole merger ends up in a new larger black hole. When two neutron stars merge, quite a lot squirts out, and the squirt will be very rich in neutrons, plus some iron from the surface of the neutron star(s). Again, these produce the right mix for the r (and perhaps p) process, and as I write, is considered perhaps the more promising candidate. In particular, LIGO's 2017 detection of gravity waves from the merger of two neutron stars is especially significant since it also produced a gamma-ray burst. Even though these neutron star mergers are very rare compared with other supernova explosions, they are now thought to produce a substantial portion of all the r-process atomic nuclei we observe, and probably most of the atoms heavier than lead.

The s-process is also interesting. It happens toward the end of helium-burning in red giants and supergiants of masses from 1 to 8 or so solar masses, including, that is, our own Sun, 5 billion years from now. The trick is that the star mixes some N-14 from the hydrogen-burning zone into the helium-burning zone, where the N-14 is converted to neon-22. One more helium nucleus zaps the Ne-22 and knocks off a neutron. Carbon-13 (also made in hydrogen-burning) will act the same way. The liberated neutrons wander around until they meet some fairly heavy nucleus, which then captures them. And, *voila*, *s*-process! Of course, this works (like the CNO cycle) only in stars that already had some heavy elements when they formed. In contrast, the *r*-process works on iron made a little earlier in the same (massive) star and so could begin in the very first generation of stars.

Thus, the third contribution of low-mass stars to nucleosynthesis, besides carbon, nitrogen, and oxygen (and iron when white dwarfs explode), is the shedding of *s*-process products in their winds and planetary nebulae. One such product, technetium, has no stable isotopes and lives only a million years. It was the discovery of Tc in a few highly evolved stars by Merrill (1952) that showed conclusively that complex nuclear reactions are going on right in front of our eyes (or telescopes).

A Few Loose Ends

So far, so good. The Big Bang made hydrogen and helium. Elements from carbon to zinc arise in peaceful nuclear reactions, mostly in massive stars. And the heaviest ones come from the *s*-, *r*-, and *p*-processes acting on iron, cobalt, nickel, and so forth that are already present.

Are we home clear? Almost, apart from some of the least massive elements in Figure 3.1. The Big Bang left behind a little bit of lithium-7 (perhaps 1% of what we see in young stars). But nowhere in the preceding paragraphs have we yet found processes that make any beryllium or boron, or the rest of the lithium. There is a good reason for this. All three are very fragile. Inside a star, they will quickly burn to helium and other sturdier elements. Like deuterium, they are destroyed in stars, not made there (except maybe lithium, which is strangely common in a very few red giants).

B^2FH attributed lithium, beryllium, and boron to an unknown *x*-process. We are now reasonably sure that they are by-products of interstellar traffic accidents. The space between the stars is not empty. Rather, it is filled with diffuse interstellar gas (see Chapter 5). In addition, **cosmic rays** fill the galaxy. These are particles, mostly protons, but also nuclei of all the other elements, moving very close to the speed of light and so carrying large amounts of energy (up to as much as a well thrown baseball, in a single particle). Cosmic rays ultimately derive their energy from supernovae, though some of the details are fuzzy (see Chapter 4),

From time to time, a cosmic ray hits an interstellar atom and busts it up. The process is called **spallation**, and when the victim is a carbon or oxygen nucleus, lithium, beryllium, and boron are among the products. We know this happens because the cosmic rays themselves, when they get to us, contain large excesses of Li, Be, and B, made by the converse process of a cosmic ray C or O being zapped by an interstellar proton and losing portions of its anatomy in the collision. Some other kinds of atoms, including fluorine and sodium, are more common in cosmic rays than elsewhere and must also be partly made by spallation.

You may never have thought much about these light elements, but lithium is used in many batteries, and in treatment of some illnesses, beryllium is a light metal employed in aircraft, and one boron compound, borax, is a traditional cleaning product.

We have now examined every element and isotope, at least briefly, and looked at most of the sites where nuclear reactions ought to occur. Nearly everything seems to be coming out even. One further site must be relevant. Minor explosions — ones that don't destroy the star — occur in some stellar pairs, when a normal (main sequence) star drizzles hydrogen from its atmosphere onto a white dwarf companion. The hydrogen builds up for a hundred thousand years or so, and then (because it is degenerate) blows up in hours. The observed phenomenon is called a **nova**, and the dominant reaction is a hot, fast CNO tri-cycle (see Figure 3.13). Isotopes more common in nova ejecta than in normal hydrogen-burning should include carbon-13, nitrogen-15, oxygen-17, and neon-21.

Another possible nova product is aluminum-26. You cannot go to the drug store and buy a box of ^{26}Al. It decays to magnesium-26 in 720,000 years, so the solar system supply is long since gone. But we know it is present in the interstellar medium (because we see the radiation emitted during the decay — another proof that nucleosynthesis is an everyday affair). And, second, we know it was present when the asteroids and meteorites formed, because we find the product, ^{26}Mg, in strange places. ^{26}Al decay was, therefore, a source of heat in young solar system objects, and it is generally blamed for having melted the asteroids that later broke up to make meteorites. This is why astronomers are interested in where it was made. Novae, supernovae, and the same stars that make *s*-process elements are all possibilities. Each has both strong proponents and vigorous opponents in the community.

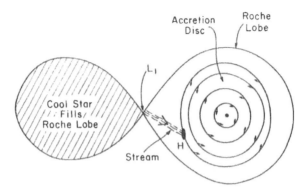

Figure 3.13. Anatomy of a nova. Material from a cool star flows in a stream that hits a disk (at H) orbiting a white dwarf (the small black dot at the center of the nearly circular accretion disk pattern). Gas in turn flows from the disk onto the white dwarf until enough has built up so that the bottom of the hydrogen-rich accreted layer is degenerate. The accreted material burns explosively and is blown back off the white dwarf and into space. The name *cataclysmic variables,* for novae and related binaries, was coined by Cecilia and Sergei Gaposchkin. Gas will always flow to the second star when the first one is larger than a certain critical volume called the *Roche lobe.* The gas flows most readily through the first Lagrangian point, L_1. The lobe and point are named for the French mathematicians who long ago studied the gravitational forces of pairs of stars or other masses.

Galactic Chemical Evolution

The last two sections described the origin of the individual elements and isotopes in some detail. But obviously this is not enough! We must get them into the galaxies, stars, and planets that we see today at the right times for the solar system to form 4.6 billion years ago, with the composition it has, or we won't be here and all the previous works would have been wasted.

The picture of the gradual increase in the abundances of the elements beyond hydrogen and helium and their sharing out among stars and gas has to be painted with a much broader brush. Three problems contribute to smearing out our image of galactic chemical evolution. One is fairly obvious and also has a fairly satisfactory solution. The others are both less obvious and less satisfactorily dealt with. Nevertheless, we try both because astronomers (and other scientists) always want to know everything about everything and because the number of stars with enough heavy elements to have Earth-like planets depends on the progress of chemical evolution.

The first difficulty is that the Milky Way contains about 200 billion stars. As you might guess, the largest computers in the world, even the largest you can imagine, will not be able to keep track of 200 billion anythings. The current record is more like one billion, and they are not stars, but elements of dark matter that go into forming proto-galaxies (Chapter 2). For decades, "too many stars" was thought to be a barrier to the serious study of galactic evolution. The problem was solved by Beatrice Muriel Hill Tinsley in her 1967 Ph.D. dissertation at the University of Texas, Austin (remember, I told you when we met Cecilia Payne that there was going to be one more spectacular thesis!). Her assumptions and approximations are mentioned back in the "History" section.

Amazingly, the models she calculated (Tinsley, 1968, 1980) look almost like real galaxies. Tragically, she died of malignant melanoma in 1981, at the age of 40, and so never knew about the second problem, though her solution to the third anticipates the modern one.

Problem two is that galaxies never, or almost never, began with a blob of gas (and dust after the first stellar generation) containing all the material that it might find and need for the rest of its life. Instead, the early stages are those of Chapter 2. Proto-galaxies of dark matter gradually gather together. Then the surrounding additional dark matter and gas flow in, and begin making Population III (meaning no heavy elements) stars. Mergers occur, more gas flows in, some gas probably gets kicked out, taking newly synthesized heavy elements with it. And so on down to the present (see Chapter 2).

Thus, vital early stages not only occurred before anyone was around to look, but also occurred in entities that no longer exist. This isn't quite hopeless, because the oldest stars in our own galaxy probably preserve the products of that first generation of stars (Frebel, 2015; Frebel and Norris, 2015). The oldest stars found so far actually have less than a millionth of the iron content found in our Sun, though carbon and oxygen are "less deficient," saying something about what that very first generation of stars did. That is, those stars tossed out heavy elements all right, but not the same mix we see being synthesized today. There is even one dwarf galaxy that looks like it could be a "last rose of summer," a relic of the units that made the Milky Way, still left in orbit around us, and made of (very few!) stars each of which perhaps has heavy elements from only one supernova. Its name is Segue 1, for the survey that found it. A major uncertainty at this level is the typical mass of a Population III star (Frebel and Norris, 2015; Somerville and Dave, 2015).

The third problem comes about because we know less about what a gas cloud wants to do than the cloud itself knows, whether it has enough material for only a few stars or for the biggest star clusters we see. The basic processes of galactic evolution are easy to list. Interstellar gas and dust turn into stars (see Chapter 5). Nuclear reactions in stars build heavy elements and some of the products are blown back out to mix with remaining, interstellar stuff. Fresh gas can flow into the galaxy or winds can expel the enriched gas. And gas of different compositions can flow from one place to another inside a galaxy. Each process is governed by ordinary laws of physics and has a unique outcome. That is, for instance, a particular gas cloud will make a definite number of stars of particular masses at some time. But even if you tell a theoretical astrophysicist all about the cloud — its distribution of density, temperature, magnetic field, and so forth — he will not be able to tell you in return just when it will form stars, or how many of each mass, or what percentage of them will be in binary pairs. The process is just too messy to calculate. Similar situations occur in other sciences. Consider weather forecasting. No unknown physics is involved. But, even with a perfect picture of where the clouds are now, which way the wind is blowing in each place, how the ocean currents flow, and so on, you don't plan an outdoor party two weeks in advance on the basis of weather reports. The inherent uncertainty of the situation is reflected in the forecasts themselves: "20% probability of rain," "sunny periods," "rain possibly turning to sleet."

The incalculable items we need for galactic chemical evolution are: (a) the total rate of star formation — in solar masses per year, as a function of time, place, and gas composition in the galaxy; (b) the *IMF* — the relative numbers of stars of each mass formed (which will also vary with time, place, and composition); (c) rates of gas flow into and out of whatever region you are considering, plus the composition of the gas involved; and (d) how fast enriched material gets mixed with its surroundings.

The standard way of cheating is called an adjustable parameter. Consider the rate of star formation. Choose a likely value, for instance, the current one in our galaxy (a few solar masses per year). Use it in your model and see if you like the answer. If you don't, vary the parameter called "star formation rate" (SFR) until you do like the answer. And, if the rate you end up liking is not too unreasonable, you have probably learned something. Parameters representing gas inflow and outflow, the stellar mass distribution, and the rest are treated similarly, one at a time.

Simple Models and the G-Dwarf Problem

Suppose, magically, you had a chance to ask a scientist living in a distant galaxy just one question and get an answer. What would it be? Perhaps (if you are politically inclined), "Do you have wars?" If medically disposed, "Can you live forever?" If theologically minded, "Do you acknowledge some superior power?" Trimble, the astronomer, would ask, "Do you guys have a G-dwarf problem?" So, let me explain what this means and why it has driven many of the more elaborate models of chemical evolution for the last 40 years or so.

Start with homogenous, instantaneous recycling, constant percentages of stars of various masses in a closed box, with nothing added or subtracted except the gradual transformation of H and He to heavy elements (in a constant mix, like solar abundances). For any choice of SFR, one easily calculates how many long-lived stars of each metal abundance should be left from all previous star formation. G-dwarfs (like the Sun) are a good choice because they can live for about 10 billion years and are bright enough to count in a volume of a million cubic parsecs or so. And no matter who does the calculating or the counting, these simple models say there should be many more metal-poor stars around than there are. Figure 3.14 (from Trimble, 1975) shows the problem: the dashed line is the calculation; the various letters are data from several different astronomers, and at the low-metal (left) side they all fall well below the prediction.

Can we fix this? Yes, of course! Let the IMF change with time, so that very few low-mass, long-lived stars were born in the distant past. Of course, there will be very few left! Give up homogeneity, so that the heavy elements blown out by each supernova lurk fairly close and there is more star formation in these metal-rich regions. Let gas of various compositions flow into and out of the region you are looking at, forming stars as it comes or goes. And there is some theoretical justification for any or all of these, because extra metals means easier gas-cooling and so more vigorous star formation.

Is it proper to care about theoretical justification, if the data are telling us something specific? Actually yes. Sir Arthur Eddington, who was among the very first to realize, in the 1920s, that stars must live on what we now call fusion reactions, said he refused to believe an observation until it was confirmed by theory. What he meant was that he was reluctant to accept things until he understood them. He also persuaded Cecilia Payne to go from Cambridge (UK) to Cambridge (MA, Harvard) so she

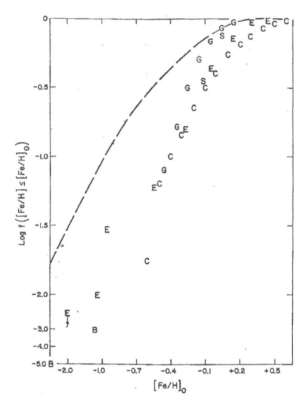

Figure 3.14. Models predict (dashed line) that a substantial fractions of stars in the galaxy should have much lower iron (Fe) abundance than the Sun. Both axes are on a logarithmic scale relative to the Sun-like stars with its abundance of heavy elements. For example, almost every 10th star should have less than 1/10th of the Sun's heavy elements. But these predictions are in stark contrast to the observations, shown by letters, which find far fewer metal-poor stars. This graphically summarizes the "G-Dwarf problem".

could get a Ph.D. He therefore counts as a good guy in the pantheon of 20th century astronomy.

The current generation of models of chemical evolution (Nomoto et al., 2013) makes use of all the flexibility just mentioned and can, thereby, account for what we see in the different parts of the Milky Way, in other galaxies, in intergalactic gas near quasi-stellar objects (quasars), and elsewhere. The most important input processes are Big Bang nucleosynthesis, winds from asymptotic giant branch stars and from novae, and the different mixes of heavy elements that come from three basic types of thermonuclear and gravitational collapse supernovae. Still not well-treated, however, are the effects of chemical differences from one place to another in the interstellar gas that forms stars, binary stars (your author's own particular pet subject), and supernovae of rapidly rotating stars.

Habitable Planets and the Future

It has been 25 years since the discovery of the first sun-like star with a planet. Much surprise was occasioned because the resemblance to the solar system was nil. Yes, the planet, 51 Peg b, had a mass like our Jupiter (the only sort that could have been detected in those days), but it was very much closer to its star than Mercury is to the Sun. Dozens of these "hot Jupiters" have turned up since, but also many other combinations of numbers of planets per star (l to 8), masses, distances from their hosts, and nonsolar host types. Technology is rapidly closing in on masses like Earth and places where water can be a liquid (Chapter 6).

For a while, it, looked like only stars more metal-rich than the Sun were likely to have planets (and this remains true for hot Jupiters). But it is not the case for smaller planet masses like Earth. Habitable planets are, therefore, likely to be found through much of the volume of the Milky Way, though probably not into the halo, nor extremely close to our central black hole, where sporadic blasts of X-rays and gamma rays are likely to erode both planetary atmospheres and the folks who were counting on breathing those atmospheres.

While the chemical (and other) evolution of the Milky Way and its kin is exceedingly complex, with violations of all the assumptions of the original Tinsley model, we also see occasional purer cases, like Segue 1 just mentioned and a galaxy about 10 billion light years away with cool gas flowing peacefully inward along a filament of denser-than-average dark matter (Chang, 2015; Martin *et al.* 2015).

At least two things are going on now and will continue into the foreseeable future: First, the processes that have turned a nearly homogeneous Universe of hydrogen and helium into all we see today, and, second, the efforts of whole cohorts of astronomers, physicists, and all to understand the processes and products. The Universe of 10 billion years into the future will be no more recognizable as ours than the Universe of 10 billion years in the past was.

References

Bethe, H. A. 1939. Energy production in stars. *Physical Review* 55: 434–456.

Bethe, H. A., and Critchfield, C. L. 1938. The formation of deuterons by proton combination. *Physical Review* 54: 248–254.

Burbidge, E. M., Burbidge, G. R., Fowler, W. A., and Hoyle, F. 1957. Synthesis of the elements in stars. *Reviews of Modem Physics* 29: 547–650.

Cameron, A.G.W. 1957. Chalk River Report CRL-41 and nuclear reactions in stars and nucleogenesis. *Publications of the Astronomical Society of the Pacific* 69: 201–222.

Chandrasekhar, S. 1931. The maximum mass of ideal white dwarfs. *Astrophysical Journal* 74: 81–82.

Chandrasekhar, S. 1935. The highly collapsed configurations of a stellar mass. *Monthly Notices of the Royal Astronomical Society* 95: 207–225.

Chang, S. 2015. Astronomers observe a nascent galaxy stuck to the cosmic web. *Physics Today* 14–15.

Frebel, A. and Norris, J. E. 2015. Near-field cosmology with extremely metal-poor stars. *Annual Reviews of Astronomy and Astrophysics* 53: 631–688.

Frebel, A. 2015. *Searching for the Oldest Stars* (Princeton University Press).

Galli, D., and Palla, F. 2013. The dawn of chemistry. *Annual Reviews of Astronomy and Astrophysics* 51: 163–206.

Gamow, G. 1949. On relativistic cosmogony. *Reviews of Modern Physics* 21: 367–373.

Martin, D.C. *et al.* 2015. *Nature* 524: 192–195.

Merrill, P. 1952. Technetium in the stars. *Science* 115: 484.

Nomoto, K., Kabayashi, C., and Tominaga, N. 2013. Nucleosyntheiss in stars and the chemical enrichment of galaxies. *Annual Reviews of Astronomy and Astrophysics* 51: 457–509.

Payne, C. H. 1925. *Stellar Astrosphysics* (Cambridge, UK: Heffer & Sons).

Scerri, E. 2007. *The Periodic Table* (Oxford University Press).

Somerville, R. S., and Dave, R. 2015. Physical models of galaxy formation in a cosmological framework. *Annual Reviews of Astronomy and Astrophysics* 53: 051–114.

Thielmann, F., Eichler, M., Panov, I., and Wehmeyer, B. 2017. Neutron star mergers and nucleosynthesis of heavy elements. *Annual Review of Nuclear and Particle Science* 67: 253–274.

Tinsley, B. M. 1968. Evolution of the stars and gas in galaxies. *Astrophysical Journal* 151: 547–565.

Tinsley, B. M. 1980. Evolution of the stars and gas in galaxies. *Fundamentals of Cosmic Physics* 5: 287–388.

Trimble, V. 1975. Origin and abundances of the chemical elements. *Reviews of Modern Physics* 47: 877–976.

Trimble, V. 1991. Origin and abundances of the chemical elements revisited. *Astronomy and Astrophysics Reviews* 3: 1–46.

Trimble, V. 2010. The origins and abundances of the chemical elements before 1957: From Prout's Hypothesis to Pasadena. *European Physical Journal H* 35: 89–109.

Chapter 4

Stellar Explosions, Neutron Stars, and Black Holes

Alexei V. Filippenko

Introduction

Our Solar System formed about 4.6 billion years ago from a **nebula**, a giant cloud of gas and dust in the cosmos (see Chapter 5 for details of star formation). Since its birth, our Sun, a typical star, has been generating energy through the **nuclear fusion** of hydrogen to helium (see Chapter 3). Two neutrons and two protons tightly bound in a helium nucleus have slightly less mass than the four original protons (hydrogen nuclei), and this mass deficit (m) is converted to energy (E) according to Albert Einstein's famous equation, $E = mc^2$, where c is the speed of light in a vacuum. In our Sun's case, about 700 million tons of hydrogen are fused each second, at a temperature of about 15 million **kelvins** (K), but the fuel supply is vast; the Sun will continue this process for another 5 billion years, increasing in brightness only slightly (except near the end) as the composition of its interior gradually changes. This **main-sequence** phase of a star lasts as long as there is hydrogen in its **core**.

Eventually, however, the core (about 10% of the total mass) consists almost entirely of helium, which requires much higher temperatures for fusion into heavier elements. As the inert core loses heat, it contracts because of the pull of gravity; consequently, energy is liberated (just as a dropped ball picks up speed). Half of the energy escapes from the star, but the other half heats the core and the surrounding layer of hydrogen; the inner part of the hydrogen shell therefore continues to fuse into helium, but at an accelerated rate, and thereby increases the mass of the helium core. The excess radiation produced by the core contraction and accelerated fusion causes the envelope of the star to expand by an enormous factor, and the star becomes a luminous **red giant** with a relatively cool surface. When our Sun eventually goes through this phase 5–6 billion years from now, its diameter will swell to about half the orbit of Mercury. The Sun will become so bright that Earth's surface will literally be fried, and all life will certainly be destroyed.

If the star's initial mass is at least half the Sun's mass, 0.5 **solar masses**, the temperature of the core grows until it reaches about 100 million K, at which point helium nuclei begin to fuse into carbon and oxygen (see Chapter 3). This phase is relatively short-lived, however, because the amount of energy liberated per **nuclear reaction** is much less than that from hydrogen fusion, and the star is much brighter. After roughly 1 billion years, the core consists of carbon and oxygen nuclei, which are not able to fuse to heavier elements because the temperature is too low. As was previously the case with the helium core, the carbon–oxygen core gravitationally contracts, becoming hotter and releasing energy. Helium and hydrogen fusion proceed more vigorously in shells surrounding the core, and the star swells up to form an even larger red giant.

At this point, the star's outer layers become unstable; they can be expelled rather gently through **stellar winds** and a series of "cosmic burps" to form a **planetary nebula** — a balloon of hot gas like the famous Helix nebula in the constellation Aquarius (see Figure 4.1). The gas is **ionized** by the ultraviolet (UV) light emitted by the surface of the star seen at the center of the Helix which is very hot, because what is now the central star used to be in the hot interior of the red giant. As electrons collide with positive ions and also recombine with them, the gas emits light and glows incandescently, often having the appearance of a ring or a disk in projection (hence the name "planetary nebula," but this structure has nothing to do with the real planets). The nebula is enriched in helium, carbon, nitrogen (produced as a consequence of hydrogen burning at high temperatures); some oxygen, and trace quantities of heavier elements created by the capture of neutrons (see Chapter 3). The heavy elements subsequently become part of the **interstellar medium** the gas, and dust between the stars.

Besides the planetary nebula, all that remains of the original star is the relatively small core seen at the center of the nebula. This star

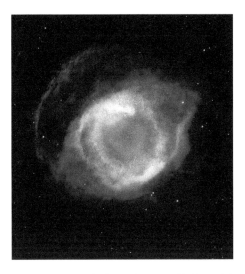

Figure 4.1. The famous planetary nebula in the constellation of Aquarius, nick-named the "Helix". The expanding shells of gas, formerly the outer layers of the dying red giant star, are heated to emit colorfully by the high-energy photons emanating from the hot white dwarf (tiny dot in the center of the image). *Source*: NASA/Hubble Space Telescope.

gradually contracts to form a dense **white dwarf**. The white dwarf consists of a carbon–oxygen core and a helium shell, the products of nuclear burning that sustained the star's power output over billions of years. Our Sun will become a white dwarf in 6–7 billion years. Slightly more than half the Sun's current mass will be compressed into a sphere roughly the size of Earth, at a *density* (mass/volume) so high that a tablespoon of the material will weigh several tons. It will be supported against the enormous pull of gravity by a purely **quantum-mechanical** pressure known as **electron degeneracy**. "Degenerate electrons" are not at all morally reprehensible; they just behave in a way that is highly unusual in comparison with classical low-density matter such as wood and bricks (see Chapter 3).

Having no new source of energy, the atomic nuclei in the white dwarf will gradually cool down while being supported by the electron degeneracy pressure. The white dwarf will slowly fade and become a dark rock hurtling through space, essentially forever. Thus, despite a burst of glory in old age manifested by the red giant and planetary nebula phases, the Sun's death will be relatively quiescent. This is the case for most stars; indeed, all single stars initially less massive than about 8 solar masses end their lives as carbon–oxygen white dwarfs, and at least some in the range 8 to 10 solar masses probably become oxygen–neon–magnesium white dwarfs. The same is true of stars in widely spaced binary systems in which transfer of substantial amounts of gas from one star to the other cannot occur.

Stellar Explosions — Celestial Fireworks

In some cases, stars can go out with a bang; they undergo tremendous explosions among the most violent of any since the **Big Bang** itself (see Chapter 1). Some **supernovae** brighten by a factor of 10^{12} to 10^{13}, as compared with 10^2 to 10^6 in an ordinary **nova**, which involves energetic processes (like nuclear fusion) only at the surface of a white dwarf. Supernovae are exciting and fun to watch (as are most bombs when not used for destructive purposes — who does not enjoy a good fireworks show?). For a few weeks, the visual brightness of a supernova can rival that of an entire small **galaxy** containing 10 billion stars. The outer layers of the gas are heated to temperatures of several hundred thousand K and are ejected at speeds up to one-tenth the speed of light (0.1c). A relatively recent, well-studied example is SN 1987A, the first supernova discovered in 1987. This object (see Figure 4.2), located in the **Large Magellanic Cloud (LMC)** only 170,000 **lightyears** away, was the brightest supernova to grace our skies since Kepler's supernova of 1604.

Supernovae are incredibly important and interesting objects. Some supernovae produce compact remnants called **neutron stars**, in which a mass equal to that of roughly 500,000 Earths (1.4 solar masses) is squeezed into a sphere whose diameter is less than that of a large city such as Los Angeles. This form of highly compressed matter cannot be created in the laboratories on Earth, but nature has done it for us, which gives us the opportunity to indirectly study the properties of the compressed matter. Some supernovae may form **black holes** — regions in which matter is compressed so much that the local gravitational field permits nothing, not even light, to escape.

Supernovae send **shock waves** (pressure jumps produced by supersonic motion) through the interstellar medium, heat the tenuous gases to millions of K, and affect the structure of galaxies. Through compression, shock waves can also help initiate the formation of new stars in denser clouds of gas. Thus, the death of one star can trigger the birth of another star.

As discussed later in this chapter, supernovae are useful for **cosmology**, the study of the overall structure and evolution of the Universe. The

Figure 4.2. A portion of the Large Magellanic Cloud, a nearby satellite dwarf galaxy. In the upper left, the glow of ionized hydrogen gas heated by young hot stars is called the Tarantula Nebula. Over to the right, the brightest star in the image is the post-explosion Supernova 1987A. *Source*: **European Southern Observatory.**

tremendous power of supernovae makes them visible from far away, so they are attractive tools with which we can measure the distances between galaxies. The technique is analogous to the way the human brain judges the distance of an oncoming car by estimating the apparent brightness of the headlights and comparing with how bright they would appear if very nearby.

But perhaps the most important fact about supernovae, from the human perspective, is that they create and disperse most of the heavy elements, thus providing the necessary ingredients for Earth-like planets and life. Studies have shown that the Universe started out consisting only of hydrogen, helium, and trace amounts of lithium and beryllium (see Chapter 3); heavier elements were synthesized through nuclear reactions deep inside stars, and produced, as a by-product, **electromagnetic radiation** (light). If these elements remained forever locked up within the stars, they would be of no use. Explosions are generally necessary to release them, like most of the oxygen that you breathe,

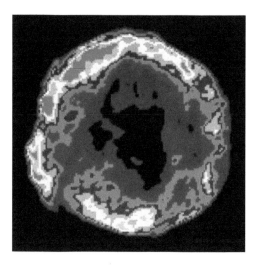

Figure 4.3. False-color image of the radio emission from high-energy electrons in the expanding supernova remnant around Tycho's Supernova, which exploded in 1572. Source: European Southern Observatory. *Source*: National Radio Astronomy Observatory.

the phosphorus in your DNA, the calcium in your bones, and the silicon in rocks. Moreover, explosions themselves (of various types) directly or indirectly produce a majority of the heaviest elements, such as the iron in your red blood cells, the gold in your jewelry, the lead that shields your body in a dentist's X-ray machine, and the uranium used in nuclear reactors (see Chapter 3 for a detailed discussion of the origin of different elements).

The chemically enriched gases are ejected into the cosmos mostly via **supernova remnants** (see Figure 4.3) which gradually spread out, as shown in Figure 4.4. Over time, the gases merge and mix with the hydrogen and helium of which our Galaxy formed, as well as with the polluted debris from the other supernovae and to a lesser extent from **kilonovae** (produced by mergers of neutron stars, as discussed later), novae, planetary nebulae, and stellar winds. When a sufficiently large quantity of gas collects in one cloud, it can begin to gravitationally contract, and stars form in the central regions (see Chapter 5). For example, we know that stars have recently been created in the Orion Nebula (see Figure 4.5 and Chapter 5). Some chemically enriched stars have rocky, Earth-like planets on which life could eventually arise (see Chapter 6). In a similar way, our Solar System formed 4.6 billion years ago out of gases that were chemically enriched, largely by many previous generations of stars. As Carl Sagan (1980) has said, "we are made of starstuff" — we owe our existence to stars and supernovae.

Stellar Explosions, Neutron Stars, and Black Holes

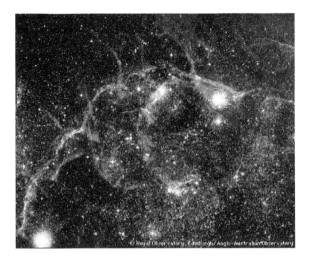

Figure 4.4. Detailed view of a portion of a supernova remnant in the constellation of Vela. To the lower right, out of this picture, one can find the star that exploded 11,000 years ago to make this expanding blast wave. *Source*: Royal Observatory Edinburgh.

Figure 4.5. Optical image of the ionized hydrogen gas in the Orion Nebula. This nebula is visible to the naked eye as the middle "star" in the sword of the constellation Orion. The power source for the glowing gas is the four recently formed hot massive stars in the bright central region of this image. *Source*: NASA/Hubble Space Telescope.

How to Find a Supernova

Astronomers are very actively studying supernovae to learn more about their physical characteristics, explosion mechanisms, and consequences.

Figure 4.6. The nearby bright barred spiral galaxy Messier 83, viewed nearly face-on. It is currently forming several new stars per year, including a few very massive ones. However we must typically wait a hundred years to observe one of these exploding as a supernova. The faint dots of light seen in the outer regions of the photograph are foreground stars in our own Milky Way Galaxy. *Source*: Anglo Australian Observatory.

However, there is a dearth of nearby objects; on average, supernovae occur only two or three times per century in large galaxies, such as the magnificent spiral M83 (see Figure 4.6), and some are hidden from our view by dust (mixed with gas) in the interstellar medium. Thus, to have a reasonable chance of discovering several supernovae in any given year, one must look at many galaxies, most of which will be distant because there are so few nearby galaxies. (It is like trying to find a two-headed snake; they exist, but are rare, so you are unlikely to see one in your own backyard.)

A careful amateur astronomer using a small telescope can make a valuable contribution to astronomy by searching for supernovae. The person can simply examine galaxies, like the one in Figure 4.6, as frequently as possible and look for any changes in appearance. The individual stars shown in Figure 4.6 are foreground stars in our own Milky Way Galaxy, but an exploding star in the distant galaxy would have a similar appearance; thus, the observer should determine whether there are any additional stars in the galaxy each time it is viewed. On the evening of March 28, 1993, for example, Francisco Garcia discovered SN 1993J (the tenth, "Jth," supernova of 1993) in this manner with his 0.25-m (10-inch) reflecting telescope near Madrid, Spain. Only 12 million lightyears away in the spiral galaxy M81, this was a very interesting object.

The champion of all visual supernova hunters is the Australian amateur astronomer Reverend Robert Evans; during the years 1981–2008 he discovered 42 supernovae. He memorized the star patterns around many hundreds of galaxies, and he could find and scan them very quickly, with a typical time of only 1 min per galaxy. Occasionally, when he saw something suspicious, he made a more detailed comparison by checking his printed chart of the galaxy. If the new object was still visible at the same location several hours later (decreasing the possibility of an asteroid), it became an excellent supernova candidate.

The same procedure can be used by amateur astronomers having small telescopes with attached cameras — new images of a galaxy can be compared with old images to find supernova candidates. This is especially easy to do with a **charge-coupled device (CCD)** rather than with a conventional photographic emulsion in the camera; CCDs are much more sensitive and provide good images of faint stars, and the data can be manipulated digitally. On the night of April 1, 1994, for example, several independent groups of amateurs used CCD cameras to discover SN 1994I in the beautiful Whirlpool galaxy M51.

One can also conduct the search semi-robotically by using a computer program to control the telescope and the CCD camera; no operator is present at the telescope during observations. The data can be farmed out to many different participants who visually compare the new and old images. In the past few decades, amateur astronomers have found many hundreds of supernovae in this manner. Among the most prominent programs is the Puckett Observatory World Supernova Search, led by Tim Puckett of Ellijay, Georgia.

Computer programs can also compare the new images of galaxies with old images of the same galaxies in search of supernovae. For instance, in the early 1990s my research group at the University of California, Berkeley used a 0.76-m (30-in.) telescope at the Leuschner Observatory (near the UC Berkeley campus) in this manner. In its two years of operation, we found seven supernovae, including SN 1994D in NGC 4526 (see Figure 4.7), about 50 million lightyears away in the Virgo cluster of galaxies.

Motivated by our success with the Leuschner prototype, my group then launched the Lick Observatory Supernova Search with the new 0.76-m Katzman Automatic Imaging Telescope (KAIT) at UC's Lick Observatory (on Mt. Hamilton, CA). During the decade 1998–2008, we led the world in discoveries of relatively nearby supernovae (typically within a few hundred million lightyears), finding more than all other supernova hunters combined

Figure 4.7. Hubble Space Telescope image of the nearly edge-on spiral galaxy NGC 4526. The bright star in the lower left corner is the supernova SN1994D discovered with the 30-inch telescope at the Leuschner Observatory. *Source*: NASA/Hubble Space Telescope.

during those years. Through the present time, we have discovered (or co-discovered) more than 1,000 supernovae, though our discovery rate is much lower than it used to be because of the changes described in the following.

Most of the CCD-based supernova searches in the 1990s and 2000s targeted individual bright galaxies, owing to the limited telescope field of view (FoV) and the relatively small detector size. In the past decade, however, new searches using wide-field cameras and large-format CCDs have revolutionized the search for transient objects throughout the sky, in an untargeted manner (that is, not confining their attention to individual nearby galaxies). For example, ASAS-SN utilizes up to 20 wide-field 14-inch telescopes (Nikon telephoto lenses) to survey the entire sky; as of early 2019, they have found about 1,000 supernovae and other transient events (such as the tidal disruption of stars when they pass near the supermassive black hole in the nuclei of galaxies). Pan-STARRS, though originally designed to search for near-Earth asteroids, has also been used to conduct a very successful supernova search. An enormously impressive set of surveys has been conducted with the 1.2-m (48-inch) Oschin Schmidt telescope at the Palomar Observatory: first the Palomar Transient Factory (PTF) with a 7.3 square degree field of view (FoV), followed by the intermediate PTF with the same FoV, and culminating with the (current) Zwicky Transient Facility with a 47 square degree FoV. This effort has so far resulted in the discovery of many thousands of supernovae.

Supernova Classification

After a supernova is discovered and confirmed, astronomers measure its **light curve** (brightness as a function of time; see Figure 4.8) and **spectrum** (brightness as a function of wavelength or color; see Figure 4.9), usually with CCDs. Both provide clues to the nature of the star and

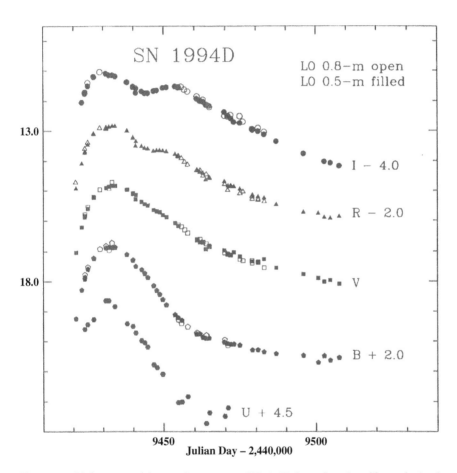

Figure 4.8. Light curves of the type Ia supernova SN 1994D through various filters, obtained by the author and collaborators with 0.5 meter and 0.8 meter robotic telescopes at Leuschner Observatory. U, near-ultraviolet; B, blue; V, visual; R, red; and I, near-infrared. The abscissa gives time in days. The ordinate units (tick marks) are magnitudes, a logarithmic measure of apparent brightness; a difference of 5 magnitudes corresponds to a brightness ratio of 100. The scale is "backward": faint stars have larger magnitudes than bright ones. All light curves except the visual one have been offset for clarity by the indicated number of magnitudes.

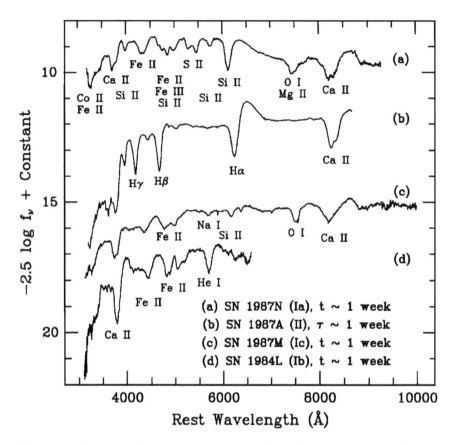

Figure 4.9. Optical spectra of supernovae on a logarithmic intensity scale, showing Type Ia, Type Ib, Type Ic, and Type II. The spectra of SN1987N and SN1987M. were obtained by the author with the 3-meter Shane reflector at Lick Observatory. These spectra were all obtained about one week after the maximum brightness of the supernova.

its explosion mechanism. Light curves are often measured through different filters transmitting blue, yellow, or red light (for example) because the temporal behavior of a supernova's brightness depends on wavelength, as may be seen in Figure 4.8. The spectrum is produced by using a prism or a **grating** (a piece of glass with many parallel grooves cut into its surface) to spread out the electromagnetic radiation into its component wavelengths from violet to red, much as raindrops turn sunlight into a rainbow. The spectrum can be used to determine the expansion velocity of the star, the chemical composition and temperature of the ejected gases, and other parameters of interest.

Supernovae come in a variety of distinct types and subtypes. The primary classification system is based on the optical spectra, with light curves providing secondary information. If the spectrum does not exhibit any **absorption lines** (dark valleys) or **emission lines** (bright peaks) corresponding to hydrogen, the supernova is called Type I. If hydrogen is present, either in emission or absorption, the supernova is called Type II. Before classifying a supernova as Type II, however, one must make sure that the hydrogen emission is associated with the exploding star rather than with the surrounding gas, as in a nebula like Orion (see Figure 4.5). This may be accomplished by determining whether the hydrogen lines appear broad. If they are broad, a wide range of **Doppler shifts** must be present, corresponding to hydrogen gas traveling at high velocities, propelled by an explosion.

Figure 4.9 illustrates the typical spectra of supernovae, roughly a week after maximum brightness. Type I (i.e., hydrogen-poor) supernovae actually come in several varieties that have subtle differences: spectra of Type Ia ("classical" Type I) supernovae exhibit a strong absorption line produced by singly ionized silicon near 6200 Å. Type Ib supernovae lack the strong silicon line but instead show lines of helium, whereas Type Ic supernovae lack strong silicon or helium lines. The large Doppler width of the lines tells us that gases move outward at speeds exceeding 5000 km/s, and in some objects reach 30,000 km/s, which reflects the rapid expansion of the supernova ejecta.

The "hydrogen versus no hydrogen" test is easy to perform; the Hα emission line of hydrogen (resulting from transitions between the third and second electron energy levels) is generally strong, and it falls in the red spectral region (about 6,500 Å) to which most CCDs are very sensitive. The test is also natural; hydrogen is by far the most abundant element in the Universe. Detailed analysis of the spectra of Type I supernovae demonstrates that hydrogen really is absent (or nearly absent) in the **progenitor** stars just before they explode, rather than somehow being hidden from view. Thus, it is reasonable to postulate that the stars giving rise to Type I and Type II supernovae are different and that their explosion mechanisms may also be distinct. As we will see, this is only partly true: while the progenitors of Type Ia supernovae differ greatly from those of Type II supernovae, the Type Ib and Ic objects are closely related to Type II supernovae.

The "I/II" (now "Ia/II") classification scheme was proposed in 1941 by Rudolph Minkowski of the Mount Wilson and Palomar Observatories.

His colleague Fritz Zwicky, an avid supernova sleuth, discovered 122 supernovae.

Explosion Mechanism — Type Ia Supernovae

Type Ia supernovae generally constitute a rather homogeneous class. At early times (within 1 month after the explosion), their spectra are dominated by lines of silicon, sulfur, calcium, oxygen, and magnesium. Although the detailed appearance of the spectrum gradually changes, at any given phase (time after explosion) the spectroscopic similarities among different Type Ia supernovae are striking. Indeed, through careful examination of the spectrum of a Type Ia supernova, it is possible to determine its phase quite accurately.

Type Ia supernovae are found in all kinds of galaxies, including ellipticals, which seem to contain primarily old stars (see Chapter 2). In spiral galaxies, the supernova rate tends to correlate with the rate of relatively recent star formation — but they show no clear preference for regions of current star formation, where the most massive stars live and die. This, together with the absence of hydrogen, suggests that we are generally witnessing the explosion of white dwarfs having ages of a few hundred million to a few billion years. The uniformity of the light curves and spectra of Type Ia supernovae further imply that the white dwarfs usually achieve a well-defined configuration before exploding.

The leading hypothesis is that a Type Ia supernova results from an uncontrolled chain of nuclear reactions in a carbon–oxygen white dwarf. This runaway is triggered when the white dwarf's mass approaches 1.4 solar masses, known as the **Chandrasekhar limit** after the Indian astrophysicist who first calculated it (see Chapter 3). Alternatively, it is possible that sub-Chandraskehar-mass white dwarfs may in some cases explode as Type Ia supernovae. Enough energy is released through nuclear fusion of light elements to blast the entire star apart; no compact remnant is created. Moreover, large quantities of **radioactive nuclei**, especially nickel, are produced through **nucleosynthesis**. The radioactive nickel subsequently decays into radioactive cobalt and finally into stable iron; this decay powers the optical light curve. (The radioactive nuclei emit gamma rays, most of which are trapped in the ejecta and are subsequently converted to optical and infrared photons.) Tycho's supernova of 1572, whose remnant is shown in Figure 4.3, appears to have been Type Ia, based on its historical light curve.

A white dwarf that eventually explodes probably grows toward the Chandrasekhar limit by **accretion** of matter from a companion star in a

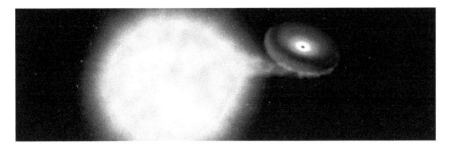

Figure 4.10. Artist's conception of a red giant star (left) in a close binary orbit with a white dwarf (right). A stream of the gas of the expanding red giant atmosphere is captured by the gravity of the white dwarf, and orbits around it in an accretion disk. Gas in this disk gradually drains onto the white dwarf surface. This accumulated mass gain could eventually push the white dwarf over its Chandrasekhar limit, causing it to collapse in a type Ia supernova explosion. *Source*: National Radio Astronomy Observatory.

binary system (see Figure 4.10), although the details of the process remain a mystery. If the companion is a reasonably normal star donating hydrogen to the white dwarf, why do we not see hydrogen in the supernova's spectrum? Moreover, to avoid the surface explosions of novae (which generally lead to a decrease in mass) and other undesirable characteristics, the white dwarf's hydrogen accretion rate must lie within a fairly narrow range of values. Another idea is that the white dwarf could be accreting helium from a star that has lost its outer envelope of hydrogen, but such systems seem very rare. In the so-called "double-degenerate" hypothesis (as opposed to the "single-degenerate" systems described before), the white dwarf reaches the Chandrasekhar limit by merging with another white dwarf; however, very few binary white dwarfs with sufficiently small separations have been found in the Milky Way Galaxy. Yet another double-degenerate possibility is the direct collision of two white dwarfs in a triple-star system.

Explosion Mechanism — Type II, Type Ib, and Type Ic Supernovae

Type II supernovae are always found in spiral or irregular galaxies, often in spiral arms and near gaseous nebulae, sites of active star formation. Thus, they are associated with massive stars, whose lives are so short that they have little chance to wander far from their birthplaces before exploding. (Although massive stars have more fuel to fuse, they do so at a prodigious rate, and use up the fuel much more quickly than do low-mass stars. For example, a 20-solar-mass star has a main-sequence lifetime of only about 10 million years; that of the Sun is 10 billion years.) The association is consistent with the observed heterogeneity of spectra and light curves of Type II supernovae; depending on their initial mass and chemical abundances, massive stars can have a wide range of sizes at the time of the explosion, and they can be embedded in gases (released by stellar winds) having vastly different densities.

Typically, the progenitors of Type II supernovae are thought to exceed 10 solar masses. In the cores of massive stars, hydrogen fuses to helium, helium to carbon and oxygen, and so on, all the way to iron (see Chapter 3). The ashes of one set of nuclear reactions become the fuel for the next set, and an onion-like radial structure develops (see Figure 3.12).

The sequence of nuclear fusion stops at iron, the most tightly bound nucleus; fusion of iron into heavier elements does not liberate energy, it requires energy. Thus, an iron core builds up in the star, which by this time is generally a red **supergiant** up to 1,000 times larger than the Sun (if our Sun were such a star, the orbit of Mars would easily be inside it). But the iron core cannot continue to grow indefinitely. When it reaches a mass comparable to the Chandrasekhar limit for white dwarfs, it is no longer able to support itself. The core begins to collapse, and electrons and protons combine to form neutrons and **neutrinos** (nearly massless, and perhaps also massless neutral particles; see Chapter 1). In the process, the core releases a tremendous amount of energy — effectively, the gravitational **binding energy** of the newly formed neutron star having a mass of 1–2 solar masses and a diameter of only 20–30 kilometers. (Matter is compressed into such a small volume that its gravitational mass is actually less, by about 0.1 solar masses, than when it was spread farther apart. This is analogous to the mass deficit of tightly bound atomic nuclei.) Most neutrinos, being very noninteracting particles, escape almost immediately.

When its density exceeds that of atomic nuclei, the central region rebounds like a basketball dropped to the floor. As it smashes into the surrounding material, it creates an outward-moving shock wave that reverses the collapse and leads to an energetic expansion. A similar bounce effect can be seen if a tennis ball is placed atop a basketball and both are dropped simultaneously. The copious neutrinos and antineutrinos (the **antiparticles** of neutrinos) emitted by the exceedingly hot (100 billion K) neutron star probably help eject the layers of gas surrounding the core; only about 1% of their energy needs to be transferred to the ejecta. During the explosion, heavy elements such as nickel, zinc, and platinum are synthesized through nuclear reactions, although the relative proportions of the elements differ from those in Type Ia supernovae.

The basic idea of core collapse in a massive progenitor star, followed by rebound, was formulated by Fritz Zwicky and his colleague Walter Baade at Caltech in the 1930s. Although the neutron had been discovered by James Chadwick only 1 year earlier, Zwicky and Baade boldly predicted that stars made almost entirely of neutrons are created by this mechanism. The density of the material would be close to that of nuclear matter — a tablespoon of it would weigh about 1 billion tons.

In some cases, the massive progenitor star may lose its outer envelope of hydrogen, or even the next layer of helium, prior to exploding. This can occur through winds emitted by the most massive stars, or via the transfer of mass to a bound companion star. If the hydrogen layer was lost, the spectrum does not show signs of that element, and the object is classified as Type Ib. If the helium layer was lost as well, then no hydrogen or helium will appear in the spectrum, and the object is classified as Type Ic. If only a small amount of hydrogen is present at the time of explosion, the object can spectroscopically appear as a Type II at early times and later transitions to a Type Ib. All of these types of objects are collectively known as "stripped-envelope supernovae."

Supernova 1987A — A Gift from the Heavens

By the early to mid-1980s, detailed theoretical studies of supernovae were being done with complex numerical codes running on large computers. Insights were gained into the nature of the progenitors, the explosion mechanisms, and explosive nucleosynthesis for both Type Ia and Type II supernovae. The predicted light curves and spectra generally agreed with the observations, at least in broad terms, but other problems remained; for example, the bounce mechanism after core collapse in a Type II or a stripped-envelope supernova could not be understood reliably, and there was no consensus on how a white dwarf actually reaches an explosive configuration to form a Type Ia supernova.

What was needed were some bright, nearby supernovae to test and refine the theories. Extensive observations throughout the electromagnetic spectrum are important because radiation at different wavelengths provides different clues to the phenomenon. Ideally, a supernova within a few

Figure 4.11. Wide-angle photograph of the southern night sky over the Paranal Observatory of the European Southern Observatory in the Chilean Andes. The four cylindrical domes on the right house the 8.5-meter mirrors of the Very Large Telescope, one of which is pointed in our direction. The bright band that arcs across the sky is the Milky Way, the stars in the the disk of our Galaxy. Directly below the Milky Way is the blue-white glow of our satellite galaxy, the Large Magellanic Cloud (LMC). Even closer to the horizon, down and left from the LMC is a second nearby dwarf galaxy, the Small Magellanic Cloud. *Source*: European Southern Observatory.

thousand lightyears could be found in the Milky Way Galaxy; after all, our Galaxy should produce at least two supernovae per century. However, most of these supernovae are distant and are obscured by patches of intervening clouds of gas and dust (galactic smog) (see Figure 4.11) because the Sun sits in the plane of the Milky Way Galaxy. The most recent easily visible supernova was observed by Johannes Kepler in 1604. One was seen in 1572 by his mentor, Tycho Brahe (see Figure 4.3), and a more recent supernova (Cassiopeia A) occurred around 1680 but was invisible (or only barely visible) to the naked eye (see Figure 3.8, and book cover).

If a nearby supernova in our galaxy is too much to hope for, the next-best alternatives are to look at the LMC and the Small Magellanic Cloud (SMC; see Figure 4.11), two dwarf satellite galaxies of the Milky Way about 170,000 and 210,000 lightyears away, respectively. The LMC, in particular, is home to a gigantic stellar nursery known as the Tarantula nebula (called 30 Doradus; see Figure 4.2). Many massive stars have formed in this region within the past 100 million years, which makes the LMC an excellent potential site for a supernova.

In 1987 our dream came true — a supernova (SN 1987A) near the Tarantula nebula was discovered by Ian Shelton at Las Campanas Observatory in Chile (operated by the Carnegie Institution of Washington) and, independently, a few hours later by several other observers in New Zealand and Australia. Shelton, an operator for the University of Toronto's

Figure 4.12. Before (Left) and After (Right) close-up images of Supernova 1987A, in the Large Magellanic Cloud. The progenitor supergiant star is indicated by the white arrow. For many weeks after the explosion, the expanding supernova blast was bright enough to be seen easily in the southern sky with the naked eye. *Source*: Anglo Australian Observatory.

0.6-m telescope, had initiated a search for **variable stars** in the LMC with a spare 0.25-m refracting telescope; on February 23, 1987, he took his first good photograph of the LMC. On February 24, he obtained his second photograph — but the wind blew shut the roll-off roof of the telescope shed, so he decided to develop the image immediately, before the night ended. Comparing the two photographs, he noticed a bright new star in the image obtained on February 24 and confirmed its presence by going outside to actually look at the LMC.

Figure 4.12 shows the "before" and "after" photographs of SN 1987A. It is not difficult to imagine the excitement of astronomers; we had been waiting nearly four centuries for such an event. The drama was conveyed to the general public — SN 1987A was the cover story for *Time* magazine (March 23, 1987 issue), major newspapers, and amateur astronomy magazines such as *Sky & Telescope* (May 1987 issue).

Word of the discovery spread fast; soon most optical and radio observatories in the southern hemisphere (from which the views of the LMC are best) devoted large amounts of time to SN 1987A. To collect X-ray and gamma-ray **photons**, to which Earth's atmosphere is largely opaque, rockets and balloons containing the appropriate detectors were launched to high altitudes. Ultraviolet photons, which are blocked by Earth's ozone layer, were observed with the NASA's orbiting International Ultraviolet Explorer. Good data at infrared wavelengths were obtained with the NASA's Kuiper Airborne Observatory; at an altitude of more than 40,000 ft, this flying Observatory was above most of the obscuring water vapor.

Testing the Theories

Optical spectra of SN 1987A revealed the presence of hydrogen, which made it a Type II supernova. A major test of the theoretical models, then, was to see whether the progenitor was a massive, evolved star. Careful examination of the existing photographs of the LMC, obtained before the explosion, verified that the progenitor, known as Sanduleak −69°202, was a supergiant star with a probable initial mass of 18–20 solar masses and an age of roughly 11 million years.

There was a twist, though — the star was a *blue* supergiant, hotter and smaller than a red supergiant. This supergiant is now thought to be a consequence of the low abundance of heavy elements in the LMC (a dwarf galaxy) in comparison with the Milky Way Galaxy (a much more massive galaxy); the structure of the atmosphere of a star depends on its **opacity**, or degree of transparency, which is related to its heavy-element content. Conditions in the deep interior of a blue supergiant are similar to those in a red supergiant, so the explosion mechanism should have been essentially the same. The relatively small size of the progenitor was also consistent with the peculiar light curves of SN 1987A — the object initially appeared unexpectedly faint because the surface area of the exploding star was small; moreover, considerable energy was required for expansion. Thus, SN 1987A confirmed that Type II supernovae arise from massive stars, but also provided valuable new information on the possible light curves and progenitor properties.

Another crucial prediction of the theory is that new elements are produced through explosive nucleosynthesis. Among the most important of these is a radioactive **isotope** of nickel, which decays with a **half-life** of a week to radioactive cobalt and subsequently (half-life 2.5 month) to stable iron. The radioactive nuclei are in excited states that emit gamma rays as they drop down to lower energy levels, in much the same way that an excited electron in an atom emits visible or infrared light when it jumps to a lower level. Moreover, as in the case of atoms, each nucleus produces a unique spectral pattern of photons. Thus, one way to confirm the synthesis of heavy elements is to search for the spectral signature of radioactive nickel or cobalt.

This was done with gamma-ray telescopes in NASA's Solar Maximum Mission satellite and in gondolas attached to large balloons flown high over Australia and the Antarctic. About 6–12 months after the discovery of SN 1987A, photons having precisely the energies corresponding to radioactive cobalt (as measured in terrestrial laboratories) were detected and confirmed. (Delay was caused by the opacity of the supernova to gamma ray photons; the ejecta had to expand and thin out before a few of the gamma rays could escape from the star and travel unimpeded toward Earth, 170,000 lightyears away.) The gamma rays had to have been emitted by nuclei synthesized during the explosion; if the cobalt were instead produced much earlier (e.g., in the gas that formed the star), it would have long ago decayed to stable iron.

The results were bolstered by infrared spectra, which showed an overabundance of nickel and cobalt, and by the optical plus infrared light curve, whose rate of decline 4 to 16 months after the explosion matched the radioactive decay rate of cobalt. (During this interval, most gamma rays remain trapped in the ejecta and convert their energy to optical and infrared photons, which escape immediately. Thus, the optical plus infrared brightness is proportional to the gamma-ray luminosity, which itself depends on the amount of remaining cobalt.) Measurements show that about 0.07 solar mass of radioactive nickel was produced by the explosion. Such an unequivocal confirmation of explosive nucleosynthesis was a major breakthrough and demonstrated beyond reasonable doubt that heavy elements are indeed cooked and dispersed by supernovae.

Neutrinos from Hell

Perhaps the most spectacular result provided by SN 1987A was the direct evidence for the formation of a neutron star. This measurement was fortuitous; it was made by separate experiments in Japan by the Kamiokande detector and in the United States by the Irvine-Michigan-Brookhaven collaboration (IMB). The experiments were designed to test certain **grand unified theories** of physics that predict that the proton is slightly unstable with a half-life exceeding 10^{30} years. The apparatus consists of an underground tank containing several thousand tons of ultrapure water and surrounded by detectors sensitive to visible light. The speed of light in water is only about 70% of its speed in vacuum (300,000 km/s). Thus, particles can exceed the local speed of light without violating the known laws of physics. If a charged particle is sent through the water at a speed faster than the local speed of light, it emits a cone of blue **Cherenkov radiation**, the electromagnetic equivalent of the sonic boom heard when an airplane exceeds the speed of sound. This radiation, along with its arrival time, can be recorded and analyzed.

How is this relevant to SN 1987A? A young neutron star is expected to be very hot —around 100 billion K. At such high temperatures, energy escapes from the star primarily in the form of neutrinos and antineutrinos. Despite not being inclined to interact with matter, a very small fraction of neutrinos and antineutrinos do interact. This is what happened in the Kamiokande and IMB water tanks — antineutrinos combined with protons to form neutrons and high-energy **positrons** (antielectrons), and the positrons sped through the water faster than the local, depressed speed of light and thereby produced Cherenkov radiation that was subsequently detected.

The Kamiokande and IMB teams each detected Cherenkov light from about 10 positrons, in agreement with expectations; roughly 30 billion neutrinos and antineutrinos passed through every square centimeter of Earth when the "flash" arrived, but their interaction probability is exceedingly small. (Indeed, only one person in a few thousand experienced a direct interaction.) Although 10 positrons may seem insignificant, the fact that they were detected by two widely separated instruments was a clear indication that a neutron star had been produced by SN 1987A. The antineutrinos arrived at the correct time (based on an extrapolation of the optical light curve to the moment of the explosion) and with the expected energies. This monumental discovery marked the birth of *extrasolar neutrino astrophysics*; previously, the only cosmic neutrinos ever detected had been from the core of the Sun.

The energy released by the core collapse of SN 1987A exceeded 10^{53} ergs — about 0.1 solar mass of material converted into pure energy according to $E = mc^2$. This is close to the energy produced in 1 s by the sum total of normal stars in the observable part of the Universe! Most (99%) of the energy of SN 1987A was carried away from the neutron star by neutrinos and antineutrinos within a few seconds after the core collapsed. About 1% was the kinetic energy of the ejecta gained largely from interactions with these elusive particles, and less than 0.01% came out at visible and infrared wavelengths. Despite being visually spectacular objects, Type II supernovae are, fundamentally, giant generators of neutrinos and antineutrinos. Their light is a sideshow.

SN 1987A provided an interesting upper limit to the **rest mass** of the electron neutrino. (Two other types are the muon and tau neutrinos; also, particles and antiparticles have the same rest mass.) If the neutrino were massless, it would be forced to travel at the speed of light, in which case all the neutrinos should have arrived simultaneously if they were emitted instantaneously. If, on the other hand, neutrinos have mass, those with higher energies move faster and arrive earlier than the low-energy neutrinos. In fact, the **antineutrinos** arrived within a time interval of about 10 seconds which set an upper limit on their mass of about 15 **electron volts (eV)**. (For comparison, in these units the mass of an electron is 511,000 eV.) However, the antineutrinos were probably not all emitted instantaneously; the hot, newly formed neutron star is opaque to neutrinos, and it takes time for them to leak out. The exact calculation is difficult, but nearly the entire observed spread in arrival times could be attributed to this effect. Thus, the 15 eV mass was by far the most stringent upper limit on the electron neutrino mass available for many years. A decade later, an independent measurement of solar neutrinos at Kamiokande established that the neutrino does, indeed, have a mass, but it is well below the SN 1987A limit, by a factor of ten or more.

Mainly as a result of SN 1987A, we now have a reasonably good understanding of the explosion mechanism of Type II (and other stripped-envelope) supernovae. Two-dimensional calculations show how matter heated by neutrino interactions is able to move outward through **convection**, which greatly increases the ease with which a successful explosion can be simulated. Abundances of elements synthesized during the explosion, together with their distribution in the ejecta, are also being determined by observations and theoretical work. Although progress has been made on detailed calculations of white dwarfs undergoing nuclear runaway, we have yet to test the theory of Type Ia supernovae as extensively as that of core-collapse supernovae; a nearby example that can be studied with a wide variety of telescopes is needed.

Gamma-Ray Bursts

Each second, somewhere in the sky there is a brief burst of **gamma rays**, as initially discovered in the 1960s with the US military's Vela spy satellites. The bursts were then abundantly confirmed by NASA's Compton Gamma-Ray Observatory, Swift satellite, Fermi Gamma-Ray Telescope, and other gamma-ray telescopes. These so-called "gamma-ray bursts" (GRBs) come in two main types: the "long" GRBs have gamma-ray emission that lasts at least 2 seconds (sometimes up to a few hundred seconds), whereas the "short" GRBs last for 2 seconds or less. Their physical nature was debated for several decades, with origins ranging from our Solar System to our Milky Way Galaxy to other galaxies at cosmological distances.

Through many dedicated studies, the counterparts of many GRBs (especially those of long duration) have been detected at other wavelengths of the electromagnetic spectrum. It is now clear that GRBs are generally located in very distant (high-**redshift**) galaxies, but they appear relatively bright because they eject beams of high-speed particles that emit light into two very narrow, oppositely directed cones, one of which is aligned along our line of sight.

Long-duration GRBs are produced when the core of a certain type of massive, rapidly rotating star collapses to form a black hole (though a neutron star might still be possible in a few cases), and the rest of the star explodes as peculiar kind of supernova (technically known as a broad-lined Type Ic supernova; neither hydrogen nor helium appear in the spectrum, and the broad lines indicate very high expansion speeds). Charged particles in the vicinity of the black hole become tremendously energized and try to escape, but they are blocked by the torus of dense material in the rotating star's equatorial plane. Instead, the path of least resistance is along the rotation axis, where the gas density is lower. The *relativistic* particles, traveling nearly at the speed of light, emerge as a jet along the rotation axis, emitting gamma rays through internal shocks. Subsequent collisions of the jet with circumstellar matter produce X-rays, optical, and other forms of electromagnetic radiation; these are seen as a much longer afterglow.

Short-duration GRBs, on the other hand, are produced when binary neutron stars spiral toward each other as **gravitational waves** are released, eventually merging and probably forming a black hole. Again, two oppositely directed jets of high-energy charged particles escape primarily along an axis, in this case perpendicular to the binary's orbital plane. If one of the jets points along our line of sight to the object, it appears much brighter than when it is seen from other, random directions. Some material explodes in other directions and glows through the decay of freshly synthesized radioactive nuclei, producing a kilonova (see page 104) because it isn't as luminous as a normal supernova but is more luminous than a bright nova. This hypothesis for the formation of short GRBs was greatly supported by the detection, on August 17, 2017, of a binary neutron-star merger in both gravitational waves and electromagnetic radiation, including the Fermi satellite discovery of an associated short GRB; see the discussion later in this chapter.

Mapping the Universe's Expansion History

In the 1990s, two teams of astronomers (I was the only person to have been a member of both teams, though at different times) were pursuing the possibility that Type Ia supernovae were sufficiently similar that we could observationally determine their intrinsic luminosity at maximum brightness and use them as "standard candles" for cosmological distance determinations. The teams started by studying the light curves of Type Ia supernovae that had been seen in nearby galaxies. These galaxies were close enough that their distances could be accurately measured by well-established methods. Knowing the distances to these supernovae, we then calculated their intrinsic peak luminosities. Although it turned out that the peak luminosities are not all the same, they are correlated with the light-curve shape (luminous supernovae have more slowly evolving light curves than intrinsically dim ones). After correction, they are remarkably uniform, so these objects are "standardizable candles" (like **Cepheid** variable stars, which exhibit a period–luminosity relation).

The power of this technique is that a simple measure of the apparent brightness and light-curve shape of a distant Type Ia supernova yields the distance of the galaxy in which it resides. Astronomers then pushed these supernova measurements out to very large distances; they were observing the Universe not as it is today, but as it was about 5 billion years ago (i.e., just before our Solar System formed).

By measuring the redshifts of these far-away galaxies with distances accurately calibrated from their Type Ia supernova peak brightness, we have the two things we need to decode the expansion history of the Universe. We can therefore measure how the rate of universal expansion changed over cosmic time. The observational results, shown in Figure 4.13, are that distant Type Ia supernovae appear fainter than expected. Their

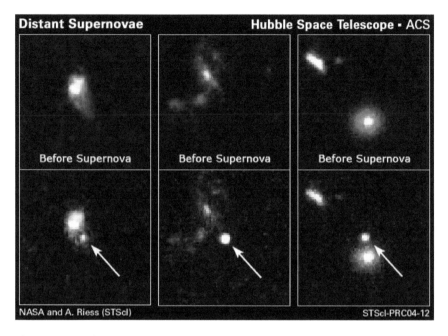

Figure 4.13. Before (top) and After (bottom) pictures of three very distant type Ia supernovae (indicated with white arrows). The brightnesses of these and other supernovae have been used to estimate the rate of acceleration of the expansion of the Universe. *Source*: NASA/Hubble Space Telescope.

inferred distances are greater than what would have been seen for these particular redshifts in any decelerating universe, or even in such an empty universe that there is no gravitational deceleration at all. How can they have reached such large distances in the given amount of time which has passed since the Big Bang?

What we found, much to our surprise, is that 4–5 billion years ago, the Universe was expanding *less* rapidly than it is now. The fact that two essentially independent teams of astronomers obtained the same result at nearly the same time made it more believable to the astronomical community. Contrary to what must happen in a universe dominated by gravity, this discovery indicated that some mysterious form of "antigravity" is now *accelerating* the expansion of the Universe. As discussed in Chapter 1, the supernova distance measurements provide evidence of a surprising **dark energy** that will cause intergalactic space to expand at an ever-faster rate… a *runaway* universe!

In 1917, Einstein came up with his **"cosmological constant,"** something of unknown physical origin that's repulsive and causes the Universe to be neither expanding nor contracting with time. There was no experimental evidence that the vacuum behaved this way; moreover, the value of the cosmological constant had to be very finely tuned to produce a perfectly static universe, and the solution was mathematically unstable in any case. After Edwin Hubble discovered that the Universe is actually expanding rather than static, the entire physical and philosophical basis for the cosmological constant vanished, and Einstein renounced it, calling it (according to astrophysicist George Gamow) his "biggest blunder." He could have predicted a dynamic universe!

About 70 years later, we have reincarnated Einstein's idea, not to make the Universe static, but to make its expansion accelerate. On the largest scales, bigger than roughly 50 million lightyears, this acceleration has dominated the motions of galaxies during the past 4-5 billion years. (Our subsequent measurements showed that the expansion decelerated during the first 9 billion years; see in the following.)

However, other things besides Einstein's cosmological constant could be causing an eventual acceleration, after an initial deceleration caused by gravity. So, we give it a more generic term, "dark energy." The cosmological constant is one possible form of dark energy. But it could be some new form of energy that permeates space rather than being a property of empty space itself. The supernova measurements, in conjunction with others, show that of the entire energy budget of the Universe, dark energy is now about 70%. With the inclusion of dark matter (25%), 95% of the Universe is still mysterious; we only know it's there, but we have not figured out its detailed physical origin or characteristics. Gradually, more evidence accumulated for this cosmic census, independent of the Type Ia supernovae. By 2011, the evidence was so overwhelming that the discovery of dark energy was acknowledged by the award of the Nobel Prize in physics to Saul Perlmutter, Brian Schmidt, and Adam Riess.

If the current acceleration of the Universe is caused by something like Einstein's cosmological constant, whose energy density remains constant, then there is a very clear prediction: at early times, the expansion should have been slowing down, and it later transitioned to acceleration. That's because galaxies started out close together, so their mutual gravitational attraction was stronger than the repulsive effect (given that there wasn't much space between them). But as they moved apart from one another, the attractive force declined, and the repulsive force increased because it's a cumulative effect. Eventually (4–5 billion years ago) repulsion started to dominate over attraction and the Universe should have transitioned from deceleration to acceleration (mathematically, this is known as a "jerk"). So, led by my former postdoc Adam Riess (who was the first to notice the evidence for acceleration, back in 1997), we used the Hubble Space Telescope to find extremely distant supernovae, up to about 10 billion lightyears away. Our analysis of the data showed that this cosmic jerk did indeed occur, 4 or 5 billion years ago.

We are continuing our efforts to more accurately measure the expansion history of the Universe, to set constraints on the physical nature of dark energy. Our latest surprise is that the current measured expansion rate (that is, the value of Hubble's constant) is 5–9% greater than expected from measurements of the **cosmic microwave background** radiation and extrapolations to the present time. The discrepancy is significant at the level of 4.4 times the measurement uncertainties, so it is being taken seriously by cosmologists. Perhaps dark energy is growing stronger with time, in which case it might eventually rip everything apart. But I think a more likely solution is that there is a new type of relativistic particle in the early Universe that affects the expansion rate or some other effect that we have not yet properly taken into account. Time will tell whether the discrepancy is real, and what far-reaching implications it has for particle physics and the cosmos!

Neutron Stars

As mentioned previously, Zwicky and Baade predicted in the 1930s that supernovae should produce neutron stars. (Some neutron stars might also form by the collapse of white dwarfs accreting matter from a companion star; these white dwarfs escape death as Type Ia supernovae for some values of accretion rate and initial mass.) Neutron stars support themselves by **neutron degeneracy** pressure, similar to the electron degeneracy pressure of white dwarfs. A few physicists, such as J. R. Oppenheimer and G. Volkoff, did theoretical studies of the possible properties of such stars, but there were no observations with which to compare. A neutron star is only about 20 to 30 km in diameter, which makes it difficult to see directly, at least in terms of radiation emitted uniformly from its surface. Very young (hot) neutron stars or neutron stars accreting matter from a bound companion star can emit X-rays, but X-ray telescopes were not in use until the 1960s and 1970s. Over four decades passed before neutron stars were actually detected — and the discovery was accidental, much like that of neutrinos from SN 1987A.

Stellar Explosions, Neutron Stars, and Black Holes

In 1967, Jocelyn Bell (now Burnell) and her doctoral thesis advisor, Antony Hewish, were using a radio telescope in Cambridge, England, to study short-timescale variations in the apparent radio brightness of astronomical objects. These are generally produced by inhomogeneities in the density of interplanetary ionized gas through which the radiation passes. Much to her surprise, Bell found that one direction of the sky seems to emit very regularly spaced pulses of radio waves. Their intensity varied considerably, but the time interval between pulses was always 1.3373011 seconds.

No known astronomical objects produced such regular, rapid pulses of radiation. After eliminating possible terrestrial sources, the astronomers briefly considered the idea of communication signals from intelligent extraterrestrial life (indeed, the object was sometimes half-jokingly called an LGM, for "little green men"). Bell, Hewish, and their collaborators, however, soon discovered three more objects having the same signature, but with pulse periods of 0.253065, 1.187911, and 1.2737635 seconds. Other similar objects (but having different periods) were subsequently found; an example of such a **pulsar** is shown in Figure 4.14. It was unlikely that different intelligent civilizations, in widely separated regions of the Milky Way Galaxy, used exactly the same method by which to communicate. Moreover, if the signal from one of the pulsars were coming from a planet, a slight periodic shift in the pulse arrival times would be expected as a result of the planet's orbital motion, but none was detected. A different explanation was needed.

The pulses could not be coming from oscillations in size of normal stars, such as in the case of Cepheid variables, since the natural periods of stars are much longer than 1 second. (The Sun, for example, experiences

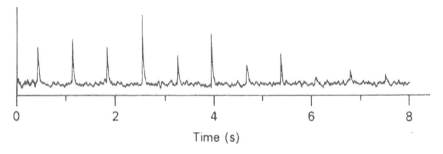

Figure 4.14. Chart record of individual pulses from PSR 0329+54, one of the first pulsars discovered. The pulse period is 0.7145 seconds. (Image courtesy Joseph H. Taylor.)

low-amplitude oscillations with a period of about 5 min.) Moreover, light from the limb (visible edge) of a star would have a longer distance to travel than light from the near side (about a 2 seconds delay for the Sun), which would thereby smear out the pulses — yet they were observed to be very narrow, with a width of only about 5% of the period. The natural oscillation period of a white dwarf, which is denser than a normal star, is 1–10 sec. — not quite fast enough for the most rapid normal pulsars (P = 0.03–0.1 sec.). Conversely, that of a neutron star is about 0.001 sec., which is too fast. Stellar oscillations of any sort therefore seemed unlikely.

Perhaps two stars were orbiting each other with a period of roughly one sec. This clearly is not the case with normal stars because they cannot get sufficiently close together. (**Kepler's third law** of orbital motion states that the square of the period of revolution is proportional to the cube of the major axis of the elliptical orbit; one finds that the stars would have to be inside each other.) The lowest orbital period of two white dwarfs is about two sec., not short enough. Two neutron stars, or a white dwarf plus neutron star pair, do not have this problem. However, the separation between the stars would be so small that they would emit gravitational waves (ripples in the curvature of **space-time**) as they orbit each other, according to Einstein's **general theory of relativity**. The system would lose energy, and the two stars would get even closer together, which would lead to a detectable shortening of the period. (This inspiral effect was actually observed recently, as described in the Black Holes section below, but not from pulsars.) Pulsars, however, are nearly perfect "clocks"; their periods seemed stable. Subsequently, more precise measurements showed that the periods were very gradually *increasing*, but certainly were not decreasing.

Having eliminated oscillation and orbital motion, astronomers were left with the possibility of rotation. A normal star is out of the question; its surface speed would have to exceed the speed of light if it were rotating once per second. A dense white dwarf is stable if rotating more slowly than a few times per second, but at shorter periods it would be torn apart by **centrifugal forces**. A rotating neutron star, on the other hand, does not have this problem; neutron stars are so dense that they can withstand rates to about 1,000 rotations per second.

Thus, astronomers concluded that pulsars are probably rotating neutron stars. If a beam of radiation were to somehow emanate from the neutron star, along an axis not coincident with the rotation axis, we would observe "pulses" when the beam intersects our line of sight once per

Stellar Explosions, Neutron Stars, and Black Holes

rotation period (or perhaps twice, if there were two oppositely directed beams whose axis is nearly perpendicular to the rotation axis). The effect would be analogous to that of a lighthouse: it is always "on," but we see it only when its beam crosses our line of sight.

What produces the beam? If neutron stars are highly magnetized, with a magnetic field pattern similar to that of Earth (a **dipole field**, which can be visualized by inserting the ends of many wires into the opposite poles of a ball) see Figure 4.15, the rapid rotation induces an electric field according to the equations of electromagnetism. This will accelerate charged particles, such as electrons and positrons, predominantly along the poles because charged particles do not easily cross magnetic field lines. They, in turn, will radiate energy along their direction of motion, which results in a beam (see Figure 4.15). Some radiated photons interact with the magnetic field and are converted into electron–positron

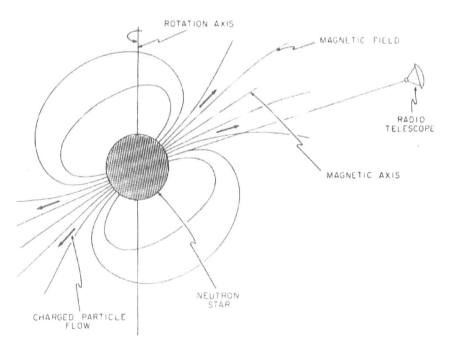

Figure 4.15. A schematic drawing of a pulsar showing the relationship between a neutron star's rotation and magnetic axes. A cone of radio emission is centered on the magnetic axis. Once per rotation, the beam sweeps across Earth (represented by a radio telescope) and we observe a pulse of radio waves. This is analogous to the lighthouse mechanism. (Diagram by David J. Helfand. Reproduced with permission from the Astronomical Society of the Pacific.)

pairs and thereby escalate the process. Although the details are complicated and controversial, many astronomers believe that pulsars "shine" because of this mechanism or one of its variants.

The rapid rotation of a neutron star is easily understood. If the progenitor star rotated appreciably (as most well-observed stars seem to do), the neutron star would naturally attain a high rotation rate as a consequence of core collapse. Objects tend to retain their total **angular momentum**, a measure of the amount of spin determined by the product of rotation rate, mass, and mass distribution. A large spinning star would have a much shorter rotation period after it shrinks, just as an ice dancer spins faster as she brings her arms closer to her body. Similarly, the magnetic field permeating a star may grow by large factors when the core collapses; the field strength is inversely proportional to the cross-sectional area of the star.

Observational Evidence

The energy emitted by a pulsar has to come from somewhere; conservation of mass plus energy is one of the fundamental laws of physics. In fact, the energy is produced at the expense of the neutron star's rotational energy; thus, the observed rotation periods of pulsars gradually increase. Eventually, the rotation is so slow that the induced electric field is too weak to support the beam-generation mechanism, and the neutron star stops shining. (Also the magnetic field decays very slowly with time, and thereby further contributes to the demise of the pulsar.) The process might take about 10 million years for a typical pulsar — far longer than the lifetime of most supernova remnants, whose gases completely merge with the interstellar medium within about 100,000 years after the explosion. Hence, it is no surprise that most pulsars are not observed to be associated with known supernova remnants.

A few key examples, reinforce this scenario. Specifically, there is a rapidly spinning pulsar (Period = 0.089 sec.) within the Vela nebula (see Figure 4.4), a supernova remnant about 20,000 years old. Even more striking is the pulsar near the center of the Crab Nebula (see Figure 4.16), the expanding remnant of a supernova that was first seen on July 4, 1054 A.D. (Although observed extensively by Asian astronomers, this supernova escaped attention throughout most of Europe and North America!) This is the youngest known pulsar and, perhaps not coincidentally, it has the most rapid rotation (30 times per second) of any normal pulsar. Although the majority of pulsars emit almost entirely at radio wavelengths, the

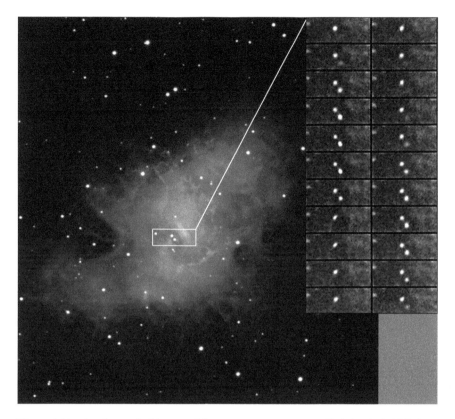

Figure 4.16. A color optical image of the supernova remnant called the Crab Nebula, a star seen by Chinese astronomers to explode in 1054 AD. The right inset shows a rapid time series of very short exposures of the central region of the Crab (indicated by the white box in the color image). This contains the Crab pulsar, which can be seen to flash on and off 30 times per second. The optical light from the pulsar produces the (rapidly varying) image, while the foreground star to its upper left does not vary its brightness. *Source*: National Optical Astronomical Observatory.

Crab pulsar is so young that it can easily be seen optically (see Figure 4.16). Having short periods and being associated with obvious supernova remnants, both the Vela and Crab pulsars played critical roles in the development of our understanding of pulsars.

The detection of antineutrinos provided convincing, but indirect, evidence that SN 1987A produced a neutron star, at least temporarily; a very hot surface had to exist from which 0.1 solar mass of gravitational binding energy was radiated in the form of neutrinos and antineutrinos. Nevertheless, to directly detect the neutron star would be exciting and

Figure 4.17. Hubble Space Telescope optical image of the supernova 1987A, taken 8 years after the explosion. The two orange rings are expanding gas which was ejected by the red giant star thousands of years before it exploded. It has now been lit up as the supernova energy reaches it. *Source*: NASA/Hubble Space Telescope.

important. After all, the neutron star may have subsequently accreted enough matter to become unstable and collapse to a black hole.

Searches for optical pulsations from SN 1987A have yielded one or two false alarms, but nothing definitive. Very little radio radiation is currently being emitted by SN 1987A. Moreover, the recent optical photographs from the Hubble Space Telescope do not show anything unusual at the center of SN 1987A. (They do, however, reveal two large rings and one smaller ring of gas that were probably produced by a wind emanating from the star before its explosion [see Figure 4.17]. These provide clues to the evolutionary history of the star.) One complication is that the ejecta of SN 1987A are still rather opaque at optical and radio wavelengths, which obscures our view toward the center. Nevertheless, if a powerful pulsar were inside, it would heat the surrounding material and cause it to glow much more brightly than what is observed. We can already rule out a source of energy in the center of SN 1987A exceeding about 1% of the luminosity of the Crab pulsar — yet, SN 1987A is far younger than the Crab pulsar, so it should be much brighter if it were born with the same rotation rate and magnetic field strength. The neutron star could also liberate substantial amounts of energy through accretion of material from its surroundings. Perhaps the neutron star in SN 1987A actually has an unexpectedly long rotational period, or there is not much gas to accrete. Although these are valid possibilities, an alternative is that the center of SN 1987A now contains a black hole.

When there are rearrangements of matter in the crust of a neutron star ("starquakes," analogous to earthquakes), a tremendous amount of magnetic energy can be released in a very short time, converted to high-speed particles and electromagnetic radiation (mostly gamma rays). Some neutron stars appear to have exceptionally strong magnetic fields (10^{15} Gauss rather than the 10^{12} Gauss typical of pulsars). Known as "**magnetars**," these objects are not yet well understood. About 30 of them have been discovered thus far.

A possibly related type of object is known as a "fast radio burst." Radio telescopes occasionally detect a single burst of radiation, having a duration of about a millisecond, whose detailed properties suggest very large distances, far beyond the Milky Way Galaxy. (The precise location of at least one fast radio burst has been found; it is in a small galaxy roughly 3 billion lightyears away.) A few such objects are known to repeat their bursts, but not in a periodic manner.

Millisecond and Binary Pulsars

There are several exotic subclasses of pulsars. First are **millisecond pulsars**: these objects rotate 50 to 1000 times per second, rather than 0.1–10 times per second as do most normal pulsars. If converted to audible sounds, the frequencies of millisecond pulsars would be similar to those of familiar notes near middle C of the musical scale. For example, PSR 1937+21 has a frequency of 642 Hz, close to E-flat in the treble clef; that of PSR 1953+29 is 163 Hz, roughly E in the bass clef. Many dozens of millisecond pulsars are now known. These objects are thought to be old neutron stars (over 100 million years old) spun up by accretion of matter from a companion star. Their emission mechanism has been rejuvenated by the very rapid rotation, despite a magnetic field that is 100 to 10,000 times weaker than that of normal pulsars.

Millisecond pulsars are incredibly stable clocks; their periods hardly change, and they pulse so many times per second that small deviations from regularity can be discerned over relatively short time intervals. This led to the discovery of tiny, cyclical variations in the pulse period of one of them. The obvious conclusion was that this particular pulsar is in orbit around at least one other object. Quantitative analysis of the data supports this hypothesis, and suggests that there are three companions whose masses resemble those of planets. These planets, however, cannot be "conventional," formed from the same gas and at the same time as the parent star; had they existed before the supernova that produced the neutron star, they would have been expelled from the system. Instead, they probably formed after the explosion, perhaps from debris remaining in the vicinity of the neutron star. It is also possible that they are the shattered remains of the companion star. Although not normal planets like Earth or those more recently discovered expolanets (see Chapter 5), these objects are very important. This was the first demonstration that planet-sized objects can form elsewhere outside our Solar System; even in what initially appear to be environments adverse to their formation.

A few pulsars are known to be in binary systems with another neutron star. The first and best known of these **binary pulsars**, discovered by graduate student Russell Hulse and his thesis advisor, Joseph H. Taylor, has an orbital period of slightly less than 8 hours — the stars are so close together that their orbit would nearly fit inside our Sun! Careful analysis of the pulse arrival times during several decades has shown that the orbital period is decreasing; the two stars are gradually spiraling toward each other. The observed inspiral rate is exactly equal to that predicted by Einstein's general theory of relativity; the system is losing energy through the emission of gravitational waves. Until recently, this provided the best test of general relativity in reasonably strong gravitational fields. The importance of the discovery was officially recognized through the award of the 1993 Nobel Prize in physics to Hulse and Taylor. (Interestingly, Antony Hewish received the 1974 Nobel Prize in physics for the discovery of pulsars, but it is a terrible injustice that Jocelyn Bell did not also. Hewish shared the prize with Sir Martin Ryle, one of the pioneers of radio astronomy).

Black Holes

The measured masses of most neutron stars in binary systems are close to 1.4 solar masses. Above a certain mass, neutron degeneracy pressure should be unable to support a neutron star against gravitational collapse. The precise value of this limit is not well known, but is thought to be in the range of 2–3 solar masses (probably closer to 2 solar masses, based on the measurements of merging neutron stars discussed in the following). Very rapid rotation (say, 1000 times/second) might increase the maximum permissible mass of a neutron star to about 4 solar masses, but this limit is not known precisely.

Calculations suggest that some stars with large initial masses (10–100 solar masses) end up with final masses that exceed this limit. They cannot lose enough matter through winds and the supernova explosion itself or through mass transfer in a binary system, so they will not end their lives as neutron stars. Similarly, a neutron star in a binary system could accrete enough material from its companion to be driven over the stability limit. In both cases, there is then no known force in the Universe that can stop the object from complete gravitational collapse. The resulting collapsed object is called a black hole.

A black hole is appropriately named — its gravitational field is so strong that *nothing*, not even light, can escape. To obtain a qualitative understanding of how this might be, consider Newton's law of universal gravitation ($F = GM_1M_2/r^2$, where G is the gravitational constant) applied to a ball of mass M_2 on the surface of Earth of mass M_1, a distance r (the radius) from the center of Earth. The force of gravity holds the ball to Earth's surface, and the ball must be thrown with a certain minimum initial speed, the **escape velocity** (about 11 km/s, neglecting air resistance), to fly completely away from Earth. Now, suppose Earth were compressed to a sphere having one-half its current radius, while retaining all its mass. The gravitational force on the ball at the surface would be four times larger than before, and the escape velocity would be multiplied by the square root of 2, which corresponds to a new value of about 16 km/s. (The formula for the escape velocity is $v = (2GM_1/r)^{1/2}$). If Earth were further compressed to a sphere having one-fourth its true radius while retaining all of its mass, the gravitational force on the surface ball would grow by a factor of 16 and the escape velocity would increase to 22 km/s.

One can imagine this shrinking process progressing to such an extent that the escape velocity formally becomes equal to the speed of light, in which case neither the ball nor anything else (including light) can escape, and the object appears black. Indeed, such arguments were made (for light) as far back as 1783 by John Mitchell and 1795 by Pierre Simon Laplace. Although the Newtonian version of mechanics (including the law of universal gravitation) is not valid when the gravitational field becomes very large, and Einstein's general theory of relativity must instead be used, the formula relating the minimum radius to which a nonrotating object of mass M must be compressed to form a black hole is the same in both cases: $R = 2GM/c^2$. Earth's radius, for example, would have to be no larger than 0.89 cm to form a black hole; that of the Sun would be about 3 km.

The radius of a nonrotating black hole, as defined above, is known as the **Schwarzschild radius**, after Karl Schwarzschild, who in 1916 used the newly developed relativity theory to formally derive the radius. Dead stars whose mass exceeds the maximum possible mass of a neutron star must continue collapsing until they are smaller than their Schwarzschild radius; thus, they must form black holes. In Einstein's theory, gravitation is actually a curvature of space-time produced by any mass or any energy (according to $E = mc^2$), and the orbits of objects are simply their natural paths in this curved geometry. In the vicinity of a black hole, space-time is curved so highly that no possible trajectories lead out from within the Schwarzschild radius. In a sense, the black hole is a "pinched off" part of the Universe from which no information can flow to the outside world. The imaginary surface that separates the black hole from the rest of the Universe is known as the **event horizon**. This boundary has a radius equal to the Schwarzschild radius if the black hole is not rotating, but the radius can be as much as a factor of two smaller in a rapidly rotating black hole.

What is the fate of the collapsing star after its radius reaches the Schwarzschild radius? According to the equations of classical general relativity, the matter keeps falling to progressively smaller radii until it is technically a point of zero volume and infinite density known as a **singularity**. In other physical situations, however, the classical laws of physics break down when exceedingly small volumes are considered; rather, one must use the laws of quantum mechanics. These have been thoroughly tested in many ways, such as by predicting and measuring the energy levels of the hydrogen atom. The quantum world appears to avoid singular points and their associated infinite quantities; the structure of matter is instead described by probability distributions of nonzero spatial extent. Although a fully self-consistent quantum theory of gravity has not yet been developed and verified (string and M-theories being the leading contenders), most researchers believe that a successful one will show that the singularity in a black hole is not really a point of infinite density. Nevertheless, the singularity is probably very small and dense — certainly, any material object would be crushed beyond recognition within it.

Fun Facts about Black Holes

Black holes have many fascinating properties. For example, if you were to fall feet-first into a black hole, you would be stretched along the length of your body and squeezed along the width by the hole's gravitational **tidal forces**. The stretching occurs because the gravitational pull on your feet (which are closest to the black hole) significantly exceeds that on your head, and the difference increases rapidly as you approach the hole. Similarly, the squeezing is produced by the fact that all points are pulled toward the center of the black hole along radial lines; your two shoulders therefore get progressively closer together. Well before actually reaching the singularity, you would resemble a long, thin string of rubber that is being stretched from both ends.

The strength of the tidal forces is a function of the mass of the black hole. Black holes formed from individual stars have enormous tidal forces even outside the event horizon, but billion-solar-mass black holes such as those at the centers of some galaxies (see the following section) seem so benign that you would not feel anything very unusual outside the event horizon. You could fall into a black hole without initially knowing that anything was amiss — but you would inexorably be drawn toward the singularity. Similarly, the average density of a black hole (defined as the hole's mass divided by the volume enclosed by the event horizon) is proportional to the inverse square of its mass — low-mass black holes have high average densities; gigantic black holes have low *average* densities, even less than water. The singularity at the center of the black hole is believed in all cases to be extremely dense.

Another important effect is **time dilation** in the vicinity of a black hole. If you were far from a black hole and watching a friend fall in, your friend's clock would appear to run progressively more slowly as he approached the event horizon. From your perspective, time would be slowing down for your friend. Indeed, as he gets infinitesimally close to the event horizon, time slows to a halt; you never actually see him reach the event horizon because this takes an infinite amount of time from your point of view. It takes a finite (and short) time from your friend's perspective, however (this is not a method for increasing one's longevity!). On the other hand, if your friend were to approach the event horizon (always remaining outside it) and subsequently escape from the vicinity of the black hole by the appropriate use of rockets, he would have aged less than you did. Hence, this is a method for jumping into the future while aging very little, much like what happens when one travels at speeds close to that of light, according to the theory of relativity.

A related phenomenon is the redshifting of radiation emitted from the vicinity of (but outside) a black hole. In the above example, suppose your friend were emitting flashes of blue light once per second on his clock while he fell toward the black hole. Not only would you see the flashes arrive at progressively longer intervals (the result of time dilation), but the flashes would appear redder and redder — that is, the wavelength of light would be stretched. Because the energy of each photon is inversely proportional to its wavelength, photons lose energy as they climb out of the deep gravitational field surrounding a black hole. A photon that is attempting to escape from the event horizon itself is redshifted to infinite wavelength (zero energy); hence, we cannot detect it.

There is a famous theorem stating that "black holes have no hair." The gist is that black holes in equilibrium are very simple objects, completely described from the perspective of an outside observer by only three quantities: their mass, electric charge, and angular momentum. In other words, external observations of a black hole cannot reveal the identity of objects that might have been thrown into it. Any small perturbations in the event horizon produced by an object falling into a black hole are quickly erased, which leaves only the three global properties of the hole.

Classically, the mass of a black hole can never decrease, because nothing can escape from within the event horizon. We expect the mass of a black hole to grow as objects in its vicinity are swallowed. However, Stephen Hawking showed that extremely small black holes do, in fact, evaporate at an appreciable rate as a result of a quantum-mechanical process. The lower the mass of the black hole, the greater its rate of evaporation — thus, low-mass black holes actually explode with a burst of high-energy radiation when their mass approaches zero. If tiny black holes were produced during the Big Bang, for example, the ones having initial masses comparable to those of large mountains (such as Mt. Everest) on Earth would now be exploding. (Less massive ones exploded earlier.) Initially, the occasional bursts of gamma rays (GRBs) detected in the sky by gamma-ray telescopes were thought by some to be these exploding *primordial* black holes, but that possibility has been eliminated because the observed properties of the bursts are inconsistent with theoretical predictions and GRBs are explained in other ways (as was previously discussed).

Detecting Black Holes

Do we have any concrete evidence that such bizarre creatures as black holes really exist in nature? Indeed we do, as shown through multiple types of objects and observations.

The first is the category of **binary X-ray sources**. In the 1960s and 1970s, after X-ray satellites were launched above Earth's atmosphere, astronomers noticed that certain parts of the sky emit X-rays profusely. Close examination revealed that in one well-observed case (Cygnus X-1), the source of the X-rays appeared to be a bright star. That star could be orbiting around a neutron star or black hole companion, and the (unseen) companion accretes gas from the star (as in Figure 4.10). Falling into the deep gravitational field of the compact companion, the gas heats up and emits X-rays. (High-energy radiation is produced because the gas is greatly accelerated by the strong gravity, and friction heats it to millions of degrees.) Spectroscopic studies of the motion of the bright star showed that it is, in fact, bound to a companion having a mass of at least 7 solar masses, and a probable mass of 15 solar masses. If the companion were a normal star having this mass, it would be luminous and easily visible, yet no object is seen. The companion could not be a low-luminosity white dwarf or a neutron star because its mass greatly exceeds both the Chandrasekhar limit and the limiting mass of a neutron star. The only reasonable conclusion is that the companion is a black hole. It is optically invisible, but the disk of super-hot gas just outside the black hole produces the strong X-ray emission we detect.

Similar systems were subsequently found. Although many turned out to contain neutron stars (with large quantities of energy released as matter hits the surface), there are now over 20 well-established black hole candidates that are X-ray binaries. One of the best is a star known as V404 Cygni, an X-ray nova that erupted in 1989. The mass of the invisible companion is about 12 solar masses, and there are fewer uncertainties than there were in the analysis of Cygnus X-1. My students and I measured the mass of the dark companion in another X-ray nova known as GS 2000+25 and found it to be at least 5 solar masses (but likely 8–9 solar masses). This is a very probable black hole; the mass exceeds the limiting mass of all but perhaps the stiffest and most rapidly rotating neutron stars. We have also measured the masses of some other examples. In my opinion, the existence of stellar-mass black holes (5–15 solar masses) in X-ray binary systems has been confirmed with greater than 99% certainty; there are no plausible alternative explanations for the observations.

Black holes have also been found in the centers of many galaxies. The gas and stars in these galactic nuclei appear to be moving very rapidly, as though a powerful force (almost certainly gravity) is pulling on them. The only plausible explanation is a giant black hole whose mass is 10^6 to 10^9 solar masses. The nearest and best case for such a "supermassive black hole" is in the center of our own Milky Way Galaxy. Special observations that compensate for turbulence in Earth's atmosphere (a technique called "adaptive optics"), thereby producing exceptionally clear images, reveal the high-speed elliptical orbits of stars in the Galactic center (see Figure 2.6). Their measured accelerations require a central black hole of about 4 million solar masses.

The presence of supermassive black holes has been inferred in more than 100 galaxies through measurements of the high speeds of stars or rotating gas disks in their nucleus. One example is the spiral galaxy NGC 4258, 20 million lightyears away — precise radio measurements reveal a disk of gas in a high-speed circular orbit around a supermassive (about 3×10^7 solar masses) black hole.

Around 1980, astronomers used ground-based measurements to argue that the nucleus of the elliptical galaxy M87, 50 million lightyears away in the Virgo cluster of galaxies, probably harbors a black hole of at least a few billion solar masses. Data subsequently provided by the Hubble Space Telescope strongly support this conclusion, with the black hole having a likely mass of about 6 billion solar masses. In April 2019, this was spectacularly confirmed by an amazingly high-resolution radio-wavelength (3 mm) map of the M87 nucleus produced by the revolutionary new Event Horizon Telescope (EHT). EHT is actually a network of eight mm-wave antennae spread across the globe. Their signals are combined to produce a map having the same effective angular resolution as would come from one gigantic radio telescope with a diameter the size of Earth. In fact, EHT would have sufficient resolution to image an apple on the surface of the Moon (if the apple were a very powerful radio emitter!).

The EHT map of the center of M87, shown in Figure 4.18, reveals a striking ring of emission surrounding a dark hole. Radio waves are being emitted from high-energy particles traveling slightly outside the black hole's event horizon. The hole in the middle is effectively the black hole's "shadow" or "silhoutte"; no radiation can escape from a black hole. Its measured size (somewhat larger than the actual event horizon, as deduced from calculations of light paths) indicates that the mass of the black hole is 6.5 billion solar masses, consistent with the Hubble Space Telescope results.

Excellent evidence for supermassive black holes has also been found with the Japanese/USA X-ray satellite ASCA; for instance, analysis of a strong iron emission line shows that gas in the galaxy MCG-6-30-15 is

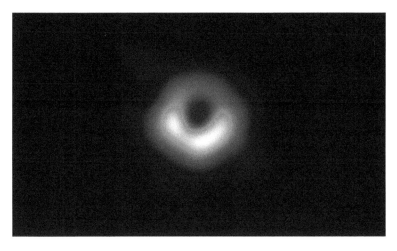

Figure 4.18   Event Horizon Telescope (EHT) image of the nucleus of the elliptical galaxy M87, showing a ring of glowing gas. The dark central region is the silhouette of a supermassive black hole (6.5 billion solar masses). (Image courtesy of EHT team.)

moving at speeds of up to 0.3c in the inner parts of an *accretion disk* surrounding the nucleus.

Interestingly, the mass of the supermassive black hole in galaxies is strongly correlated with the mass of the galactic bulge (the spherical or elliptical volume consisting of old stars orbiting the central region (see Figure 2.6)), and even more strongly correlated with the degree to which those bulge stars are compressed into a relatively small volume. These correlations provide clues to the formation and growth of supermassive black holes as galaxies evolve through time.

The specific galaxies mentioned above (NGC 4258, M87, and MCG-6-30-15) are **active galaxies** — galaxies whose nuclei emit radiation that cannot be produced by stellar processes alone. In some cases, the so-called **quasars**, the nuclei can be 10 to 1,000 times more powerful than the entire galaxy of normal stars. It has long been conjectured that supermassive black holes accreting matter from their surroundings are the central engines of active galaxies. Before being swallowed, the gas can liberate a tremendous amount of energy as it falls toward the black hole — roughly the equivalent of 10% of its rest-mass energy. This is more than 10 times more efficient than the conversion of hydrogen to helium through nuclear fusion in stars.

The existence of highly luminous quasars within the first billion years after the Big Bang demonstrates that somehow, supermassive black holes are able to form very rapidly. They then grow even more massive through accretion of matter for millions or billions of years. We do not know if they began life as small stellar-mass black holes like the ones found in X-ray binaries. Or, perhaps an entire central star cluster might have collapsed in the early stages of a galaxy's formation. But a recent breakthrough discovery has stunned astronomers into considering more massive "stellar" black holes than the 5–15 solar mass holes described before.

On September 14, 2015, the extraordinarily sensitive **Laser Interferometer Gravitational-wave Observatory (LIGO)**, in Washington and Louisiana) detected, for the first time, the gravitational waves (ripples in the fabric of space-time) produced by the merger of two black holes. These had masses of 30.6 and 35.6 Suns, and the resulting black hole was about 63 solar masses; the difference (3 solar masses) had been converted into the energy of the gravitational waves according to $E = mc^2$. They were detected by LIGO from a distance of 1.3 billion lightyears. The length of LIGO's 4-km arms changed by only about 1/1,000 of the diameter of a proton, equivalent to the nearest star other than the Sun (4.2 lightyears away) changing its distance by the width of a human hair (see Figure 4.19)! This phenomenal discovery, called GW 150914, was recognized with the 2017 Nobel Prize in physics to Rainer Weiss, Kip Thorne, and Barry Barish, though a team of more than 1,000 physicists and engineers was actually involved.

In the two years after GW 150914, LIGO detected many additional merging pairs of black holes, aided in August 2017 by the similar VIRGO detector in Italy. A surprising aspect of these detections was the high mass of each black hole in a typical merger: 20–40 solar masses, substantially larger than black holes in X-ray binary systems. How did such unusually massive stellar-mass black holes form? This is a major new puzzle in the study of black holes.

On August 17, 2017, LIGO/VIRGO witnessed the probable formation of a new black hole from the merger of two neutron stars in a binary system about 130 million lightyears away. As mentioned in the earlier section on neutron stars, this event, GW 170817, was most notable because of its additional detection via radiation throughout the electromagnetic spectrum, thereby signaling the beginning of what is now often called "multi-messenger astronomy" (although it should be noted that the detection of neutrinos and light from SN 1987A also constitutes "multi-messenger astronomy"). The Fermi gamma-ray telescope witnessed a

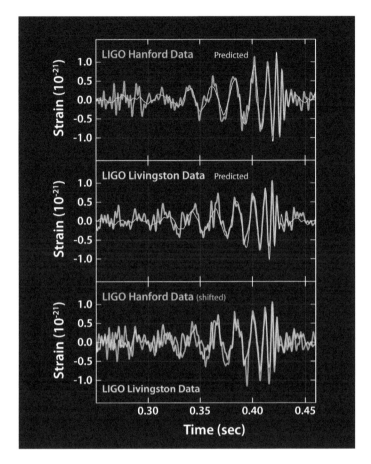

Figure 4.19 The first detection of gravitational waves. In these graphs of 0.2 seconds of data, LIGO's two detectors, in Washington (top) and Louisiana (middle), both detected the same signature of compressing and stretching spacetime. The top two panels show the actual LIGO observations with squiggly lines, along with smooth thin lines that show the theoretical predictions from Einstein's General Theory of Relativity. The oscillations due to the gravity waves speed up as the two black holes that produced them merge closer together and rapidly coalesce. The fact that the two independent detectors measured identical signals is shown in the lower panel, which compares both of their signals. This close agreement, along with the excellent match to theoretical expectations for a black-hole merger, left no doubt that gravity waves have finally been discovered. *Source*: Laser Interferometer Gravitational-wave Observatory.

short-duration GRB essentially coincident in time and space with the gravitational waves, confirming the hypothesis that such GRBs can be produced by binary neutron-star mergers, and that gravitational waves travel at the speed of light (as predicted by the general theory of relativity). Moreover, an optical/infrared counterpart of GW 170817 was discovered in the galaxy NGC 4993. Studies of the filtered light curves and spectrum of this kilonova revealed the probable presence of heavy chemical elements such as gold, silver, platinum, and the lanthanides and actinides, all of which are produced through the rapid capture of neutrons in a neutron-rich environment (not a regular supernova). This process is discussed further in Chapter 3.

Just as SN 1987A opened the observations of neutrinos from deep space, LIGO and VIRGO have now initiated the dawn of gravitational-wave astronomy. It is sure to tell us much more about the formation and growth of black holes throughout the Universe and across cosmic time.

Myths about Black Holes

Some popular myths concerning black holes should be debunked. The first is that black holes suck up everything in sight, like giant cosmic vacuum cleaners. This is not the case. A black hole's range of influence is limited; only objects in its immediate vicinity will be strongly pulled toward the hole, and even then it is possible to achieve stable (or nearly stable) orbits a safe distance away. For example, if the Sun were to somehow be transformed into a black hole (perhaps by an enormous vise), Earth's orbit would not be altered; the masses of the Sun and Earth would remain constant, as would the distance between them, so the force (according to Newton's law of gravitation) would be unchanged. Indeed, the gravitational field would remain the same everywhere outside the current radius of the Sun. Only at radii smaller than that of the Sun would the force be stronger.

Second, black holes do not form anywhere and for no apparent reason, as is sometimes implied by cartoons. They might be produced (1) by very massive stars, (2) by neutron stars accreting mass from gravitationally bound companions, (3) by neutron stars merging, (4) in the dense centers of galaxies, and (5) by inhomogeneities in the density of matter shortly after the Big Bang. In other situations, they are very difficult to make. Our Sun, for example, definitely will not turn into a black hole.

Third, travel to other Universes, or to other parts of our Universe, is probably not possible, at least for macroscopic objects. This misconception

Stellar Explosions, Neutron Stars, and Black Holes

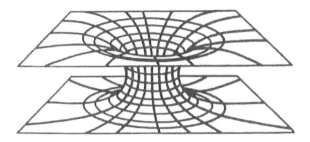

Figure 4.20. Schematic diagram of two nonrotating black holes connected by a wormhole. The curved lines denote the curvature of space near a black hole in this two-dimensional spatial "slice" of four-dimensional space-time.

arises in part from diagrams such as that shown in Figure 4.20: one black hole is connected to another black hole by a tunnel, or **wormhole** (officially called an Einstein–Rosen bridge), and it appears possible to go directly through this tunnel. However, the map is misleading; it does not adequately describe the structure of space-time inside a black hole and it applies for only an infinitesimal moment of time. More detailed analysis shows that to survive a passage through a nonrotating black hole, an object would have to exceed the speed of light, which is impossible.

The geometry of a rotating or a charged black hole is quite different, on the other hand, and travel through one at subluminal speeds initially seems feasible. However, the favorable geometry is only valid for an idealized black hole into which no material is falling; as soon as an object actually tries to traverse the wormhole, the throat closes because of the additional space-time curvature induced by the infalling material. One needs to have a very exotic form of matter to keep it open, and there is no evidence for the existence of matter having such properties, at least not in measurable quantities. (The matter must have a negative energy density in the reference frame of a beam of light passing through the wormhole.) Thus, travel via black holes to other Universes or other parts of our Universe is probably not possible, despite the allure of such a concept for science-fiction writers.

Conclusion

Astronomers are fascinated with supernovae, neutron stars, and black holes. We have learned much during the past few decades, but our understanding of the endpoints of stellar evolution is still far from complete. Stay tuned!

Further Reading

General Overview of Astronomy

Filippenko, A. V. 2006. *Understanding the Universe: An Introduction to Astronomy* (2nd Ed.) (Chantilly, VA: The Great Courses).
Pasachoff, J. M. and Filippenko, A. V. 2019. *The Cosmos: Astronomy in the New Millennium* (5th ed.) (Cambridge: Cambridge University Press).
Rees, M. 2012. *Universe: The Definitive Visual Guide* (London: Dorling Kindersley Ltd).

Stellar Evolution and White Dwarfs

Cooke, D. A. 1985. *The Life and Death of Stars* (New York: Crown).
Griffiths, M. 2012. *Planetary Nebulae and How to Observe Them* (New York: Springer).
Kippenhahn, R. 1993. *100 Billion Suns: The Birth, Life, and Death of the Stars* (Princeton: Princeton University Press).
Meadows, A. J. 1978. *Stellar Evolution* (2nd Ed.) (Oxford: Pergamon Press).
Sagan, C. 1980. *Cosmos* (New York: Random House).

Supernovae

Goldsmith, D. 1989. *Supernova!* (New York: St. Martin's Press).
Kirshner, R. P. 2002. *The Extravagant Universe: Exploding Stars, Dark Energy, and the Accelerating Cosmos* (Princeton: Princeton University Press).
Marschall, L. A. 1994. *The Supernova Story* (Princeton: Princeton University Press).
Panek, R. 2011. *The 4 Percent Universe: Dark Matter, Dark Energy, and the Race to Discover the Rest of Reality* (Boston: Houghton Mifflin).
Schilling , G. 2002. *Flash! The Hunt for the Biggest Explosions in the Universe* (Cambridge: Cambridge University Press).
Wheeler, J. C. 2007. *Cosmic Catastrophes: Exploding Stars, Black Holes, and Mapping the Universe* (2nd Ed.) (Cambridge: Cambridge University Press).

Neutron Stars and Pulsars

Bloom, J. S. 2011. *What Are Gamma-Ray Bursts?* (Princeton: Princeton University Press).
Greenstein, G. 1983. *Frozen Star* (New York: Freundlich Books).
http://www.aip.org/history/mod/pulsar/pulsar1/01.html presents the story of the discovery of the first optical pulsar told by the scientists themselves.

Black Holes

Bartusiak, M. 2017. *Einstein's Unfinished Symphony: The Story of a Gamble, Two Black Holes, and a New Age of Astronomy* (Updated Edition). (New Haven: Yale University Press).

Begelman, M. and Rees, M. 2009. *Gravity's Fatal Attraction: Black Holes in the Universe* (2nd Ed.) (Cambridge: Cambridge University Press).

Filippenko, A. V. 2011. *Black Holes Explained* (video lectures). (Chantilly, VA: The Great Courses).

Gates, E. 2010. *Einstein's Telescope: The Hunt for Dark Matter and Dark Energy in the Universe* (New York: W. W. Norton & Company).

Hawking, S. W. 1988. *A Brief History of Time* (New York: Bantam Books).

Kaufmann, W. J. III. 1973. *Relativity and Cosmology* (New York: Harper & Row).

Kaufmann, W. J. III. 1979. *Black Holes and Warped Spacetime* (New York: W. H. Freeman).

Levin, J. 2016. *Black Hole Blues and Other Songs from Outer Space* (New York: Knopf).

Melia, F. 2007. *The Galactic Supermassive Black Hole* (Princeton: Princeton University Press).

Ostriker, J. and Mitton, S. 2012. *Heart of Darkness: Unraveling the Mysteries of the Invisible Cosmos* (Princeton: Princeton University Press).

Schilling, G. 2017. *Ripples in Spacetime: Einstein, Gravitational Waves, and the Future of Astronomy* (Cambridge, MA: Harvard University Press).

Shipman, H. L. 1980. *Black Holes, Quasars and the Universe* (2nd Ed.) (Boston: Houghton Mifflin).

Thorne, K. S. 1994. *Black Holes and Time Warps: Einstein's Outrageous Legacy* (New York: W.W. Norton & Co.).

Thorne, K. S. 2014. *The Science of Interstellar* (New York: W. W. Norton & Co.).

Interesting Web Sites on the Internet

https://eventhorizontelescope.org/
https://imagine.gsfc.nasa.gov/science/objects/dwarfs1.html
http://hyperphysics.phy-astr.gsu.edu/hbase/Astro/whdwar.html
https://royalsocietypublishing.org/doi/full/10.1098/rsta.2011.0351
https://www.ligo.caltech.edu/
http://users.monash.edu.au/~johnl/StellarEvolnV1/

Chapter 5

The Origin of Stars and Planets

Fred C. Adams

Introduction

The formation of stars and planets represents one of the most fundamental problems in astrophysics. Stars provide most of the energy that is generated by the Universe and planets provide viable platforms for the emergence of life. In recent years, a lot of progress in this area has been made. In particular, we now have a fairly successful paradigm that provides the cornerstone of our current understanding of the star-formation process. Within this paradigm, the agreement between observations and theory is quite good, especially for the case of low-mass stars. We have also witnessed a revolution in the detection of exoplanets, i.e., planets orbiting other stars. This development has driven the progress in our understanding of how solar systems are made.

In essence, the process of star formation — and planet formation — boils down to a war between **entropy** and gravity. This same battleground is relevant for stellar evolution in general. In this context, "entropy" manifests itself as pressure forces and **turbulence**, although the detailed nature of turbulent motions is still under study. Loosely speaking, gravitational forces tend to pull things together, whereas entropy tends to spread things out. As this chapter illustrates, this war between gravity and entropy takes place on many different scales of size and mass. Furthermore, this battle largely determines how stars and planets form and evolve.

The chapter begins with an overview of the current theory of star formation. Stars form within **molecular cloud cores**, which subsequently collapse to produce star/disk systems. This discussion includes the **initial mass function** (IMF), the distribution of stellar masses at their births. The IMF is essential for understanding the effects of star formation on galactic evolution, galaxy formation, and other important astrophysical topics, although an *a priori* theory of the IMF remains a fundamental unresolved issue (see Chapter 3). The following section discusses **circumstellar disks**, which play an important role in the formation of both stars and planets. Several physical mechanisms have been proposed to drive the dynamical evolution of disks, including gravitational instabilities and viscous accretion. Circumstellar disks also provide the setting for planet formation, which is considered in the subsequent section. The chapter concludes with a summary and a discussion of unresolved issues.

Star Formation in Molecular Clouds

In the star-formation paradigm that has emerged in the past few decades (Shu, *et al.*, 1987; McKee and Ostriker 2007), stars form within **molecular clouds**, which are large, massive aggregations of molecular gas, typically found around the spiral arms of galaxies like our Milky Way. In fact, stars are forming today in nearby molecular clouds, which provide useful laboratories to study the star-formation process (see Figure 5.1). These clouds are much denser and much colder than the surrounding interstellar gas; they have typical number densities of approximately 100 atoms cm^{-3} and typical temperatures in the range $T \approx 10$–50 K. For comparison, note that for ordinary air at room temperature, the number density exceeds $n \approx 10^{19}$ molecules cm^{-3} and the temperature T is about 300 K. Molecular clouds are much larger than stars and typically have masses 10^4 to 10^6 times the mass of the Sun (where this **solar mass** scale is written as 1 $M_\odot$).

Stars actually form out of the collapse of cores of molecular clouds. The core regions, small subcondensations within the much larger molecular clouds, are thus the actual sites of star formation. The cores are supported by a combination of magnetic fields and turbulence, in addition to thermal gas pressure. All of these contributions act together to help support the cores against gravitational collapse. The turbulence tends to decay and leave the central regions of the cores in a more quiescent state. In addition, the magnetic fields evolve through a diffusion process, where the

Figure 5.1. A region of current star formation. Image shows a three-color composite of the sky region of M 17, an H II region excited by a cluster of young, hot stars. A large silhouette disc has been found to the southwest of the cluster center. The present image was obtained with a near-infrared camera on an 8.2-meter diameter telescope at Paranal Observatory in northern Chile. (Image Credit: ESO).

fields slowly drift outward and the inner regions of the core become increasingly centrally concentrated. Both the turbulent and magnetic contributions to the pressure support decrease with time, until thermal pressure alone supports the core against its self-gravity (at least in the central regions). At this point, the core is in an unstable state (near, but not in, equilibrium), which represents the initial conditions for the subsequent dynamic collapse. The preceding description is idealized: in practice, the central regions do not become completely divested of turbulence and magnetic fields before collapse occurs. The resulting scenario is thus somewhat more dynamic, and chaotic. In any case, the formation and early evolution of these cores represents the first stage of the star-formation process.

When a core undergoes collapse, a small pressure-supported object — the object that will become the star itself — forms at the center of the collapse flow (see Figure 5.2(a)). The cloud cores are rotating in their initial states.

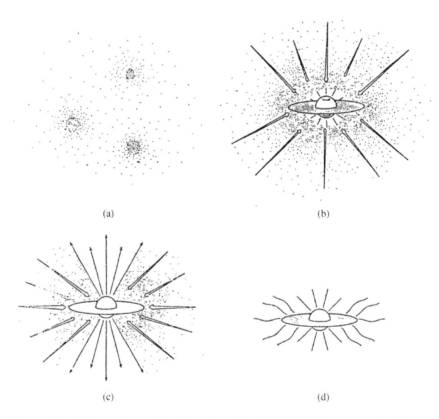

Figure 5.2. The four stages of star formation (Shu et al., 1987). Dense cores slowly condense (a), and then collapse to form a stellar-like body that is surrounded by a disk and an infalling envelope of dust and gas (b). A powerful wind emerges from the star (c) and eventually the newly formed star/disk system is revealed (d). In (b) and (c), the thicker inward arrows represent infalling material, the thinner outward arrows represent outflowing wind. In (d) the straight lines represent visible radiaiton emitted by the star, and the bent lines represent infrared radiation emitted by the disk.

Even though they are spinning slowly, because they are so large (almost a light year across), they contain a substantial amount of **angular momentum**. The infalling material with higher angular momentum collects around the forming star and creates an accompanying circumstellar disk. The presence of this disk is thus a natural consequence of the law of conservation of angular momentum. This phase of evolution, often denoted as the protostellar phase, is thus characterized by a central star and disk, surrounded by an envelope of gas and dust falling inward toward the central object (Figure 5.2(b)). The properties of this infalling structure largely determine the characteristics of the radiation that is emitted by the object during this phase of evolution; such radiation is then detected by astronomers at far away places such as Earth and allows us to see the process of star formation in action.

As a **protostar** evolves, both its mass and **luminosity** (the rate of energy generation) increase. The protostar eventually develops a strong stellar wind that breaks out through the infall at the rotational poles of the system and creates a bipolar outflow. This phase of evolution is denoted as the **bipolar outflow** phase (Figure 5.2(c), and an image of the bipolar outflow in HH30 is shown in Figure 5.3). For much of this phase, the outflow is relatively narrow in angular extent and infall takes place over most of the solid angle centered on the central star. The outflow gradually widens in angular extent, and **precesses**, and the total amount of the infalling material gradually decreases. As a result, as time goes on, the star becomes less deeply embedded within its molecular cloud core.

The outflow helps to separate the newly formed star/disk system from its parental core and the object becomes a young star. The latter stage of evolution is denoted as the T Tauri phase or the **pre-main-sequence** phase. During this stage of evolution, the system often retains its circumstellar disk (Figure 5.2(d)). Planets also form within the disk during this evolutionary epoch. Although the newly created star itself is optically visible, it does not have the right internal configuration to generate energy through nuclear fusion of hydrogen. Instead, the star generates most of its energy through gravitational contraction. As the star contracts, however, the central temperature increases until hydrogen fusion can take place. When the ignition of hydrogen occurs, the star is fully formed. The paradigm of star formation can be neatly packaged in terms of four stages, as shown in Figure 5.2.

Molecular Clouds — The Birthplace of Stars

As outlined above, molecular clouds are the birthplaces of stars. One important aspect of these clouds is that they are *not* collapsing as a whole

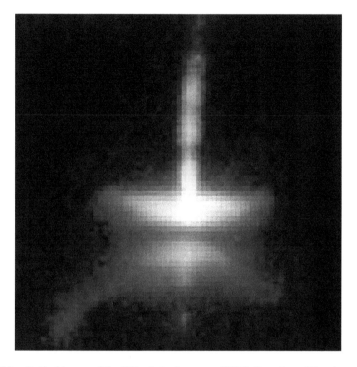

Figure 5.3. Optical image of the T Tauri star known as HH30. Our view of the circumstellar disk is nearly edge-on, so the dust grains in the disk — which presents itself as a dark horizontal band in the center of the image--absorb the light from HH30. Some light reflecting off the top and bottom of the disk is visible as flaring, mostly horizontal, bands of light above and below the dusty disk. Our special viewing orientation affords an excellent view of the bi-polar jet, which is visible in the (green) light of hot ionized gas emerging from the north and south poles of the star/disk system. *Source*: NASA/Hubble Space Telescope.

(Zuckerman and Palmer, 1974). This finding is somewhat surprising because they are so massive: the total amount of self-gravity in these cloud systems is generally much larger than the amount of thermal pressure support. In other words, the gravitational forces within the cloud are generally stronger than the ordinary thermal pressure force (see Shu *et al.*, 1987). As a result, the clouds must be supported by some additional outward force. The most viable candidates for providing this support are magnetic fields and turbulence.

Magnetic fields provide an important additional source of support and can prevent molecular clouds from collapsing. Very roughly, for a cloud of a given size and mass, a minimum magnetic field strength B is required to prevent collapse (Mouschovias and Spitzer, 1976). Observations of magnetic fields in molecular clouds show that the magnetic field strengths are generally large enough to support the clouds against gravitational collapse (Myers and Goodman, 1988; Heiles *et al.*, 1993). In particular, the field strengths are typically in the range $B = 10$–30 microgauss (these field strengths are thus about 10^5 times smaller than the strength of the magnetic field on Earth, but the fields in the clouds cover an enormously larger volume of space). In addition, we can determine the geometry of the magnetic fields in these clouds. The polarization directions of the background starlight seen through the clouds are thought to trace patterns in the magnetic field lines. Observations show that the polarization directions are generally uniform and well ordered (Goodman, 1990). We thus infer that the magnetic field itself also has a well-ordered structure. In other words, the magnetic field is not extremely tangled; it retains a more-or-less uniform component, which suggests that the field is strong enough to affect the cloud structure.

Turbulent motions are also important in opposing the gravitational collapse of molecular clouds (Zuckerman and Evans, 1974; see also the review of McKee and Ostriker, 2007). We determine the velocities of the molecules of gas along our line of sight by measuring the spread of Doppler shifts this produces in the emission lines they emit. What is usually observed is a very wide spread of wavelength shifts, corresponding to a very wide spread of gas velocities. In fact, these velocities are often much larger than those produced simply by the random thermal motions of the molecules. These indicate that on small size-scales, the gas has extra velocities that are faster than the speed of sound. This is supersonic turbulence. Moreover, the energy contained in these turbulent motions is so large that it can balance the gravitational attraction, influencing the long-term evolution of the cloud.

The formation of the substructure within molecular clouds remains an important problem. Molecular clouds have a highly complicated substructure on virtually all size scales that one can observe, and this structure is sometimes described as having a **fractal** nature (Houlahan and Scalo, 1992). In particular, these clouds do not have a single, well-defined, characteristic density; instead, the clouds have an extremely wide range of different densities. The geometrical distribution of the cloud material is also quite complicated. Cloud structures have been described in terms of "sheets," "filaments," and quasispherical "clumps" of cloud material (starting with Blitz, 1993). Moreover, the general cloud structures predicted in Zuckerman and Evans (1974) can now be observed in high spatial resolution CO maps (e.g., see Rice *et al.*, 2016 and references therein).

At present, no definitive theory exists for the formation of substructure within the molecular clouds. However, we have somes clues to this puzzle. Whenever a cloud begins to collapse, a wide spectrum of wave motions can be excited (Arons and Max, 1975), and these wave motions may provide part of the explanation for the observed clumpy structure. Numerical simulations can now follow the dynamics of this process, including the effects of gravity, turbulence, and magnetic fields (Mac Low and Klessen, 2004). On the smallest size scales, the cloud produces a wide variety of molecular cloud cores that sample a range of sizes. Although the masses of these cores are larger than the masses of the stars that they will ultimately form, they have a well-defined distribution of masses and thus provide a well-defined distribution of initial conditions for the next stage of the star-formation process.

The Formation of Molecular Cloud Cores

Molecular cloud cores represent the cloud structure on small scales of size and mass. These scales are small compared with the overall molecular cloud, but are still significantly larger than the masses of the forming stars themselves. The formation of the cores occurs through the complicated interplay among magnetic field diffusion, colliding flows, and decaying turbulence. We first consider the diffusion of magnetic fields (Mouschovias, 1976; Shu, 1983; Nakano, 1984; Lizano and Shu, 1989), which takes place as follows: the cloud is extremely lightly **ionized**; only about one particle per million carries a charge. Magnetic fields can only exert forces on charged particles. Thus, the charged particles, — the ions, — interact directly with the magnetic field, but the neutral particles (which make up most of the mass) do not. The only interaction that the neutral particles have with the magnetic field is through interactions with the ions. These interactions occur through a frictional force that, in turn, depends on the relative motion between the ions and the neutrals. As a result, ions can exert a force on the neutrals only if the two species are moving with respect to each other. The important physical implication of this situation is that the ions — and hence the magnetic field — must be slowly moving outward relative to the neutral component, which itself is slowly contracting because of gravity.

This diffusion process takes place on a time scale that is generally longer than the free-fall collapse time scale of the cloud core (the time it takes the core to collapse in the absence of any pressure forces). In the absence of turbulence, the magnetic diffusion time is roughly 10 times longer than the dynamic free-fall time for typical molecular clouds. Turbulence speeds up the process, and dominates the evolution in some regimes. Nonetheless, the processes of losing magnetic fields and turbulence (and hence pressure support of the core) provides an important bottleneck in the star-formation process. This complication keeps the efficiency of star formation fairly low in molecular clouds. In other words, only less than 1% of the molecular cloud material ends up as stars.

The mass scale defined by the molecular cloud cores is still much larger than that of the forming stars, by a factor of approximately 3–10, depending on how one defines the outer boundary of the cores. As a result, the molecular cloud as a whole cannot, by itself, determine the mass scale of the forming stars. Instead, the molecular cloud produces a wide variety of smaller core regions, which in turn provide a diverse set of initial conditions for star formation. In particular, the cores provide the starting conditions for protostellar collapse, as described in the following section.

Molecular Cloud Cores — Initial Conditions for Collapse

Stars form within the molecular cloud cores, and the observed core properties provide the initial conditions for the protostellar collapse. In the nearby molecular cloud complexes that are forming low-mass stars, the cores are observed to be slowly rotating and have a relatively constant temperature. In an idealized, but nonetheless instructive, description of the star-formation process, these cloud cores (the initial state) can be described by two physical variables: the temperature and the initial rotation rate of the core. Observations of cloud cores indicate temperatures in the range $T = 10$–35 K (starting from Myers, 1985). The measured rotation rates of these cores are extremely slow — the measured angular velocity is about $\Omega \sim 3 \times 10^{-14}$ radians/second. Thus, it takes millions of years for a cloud core to make a single complete rotation. The central regions of these cloud cores are highly concentrated, and are expected to have density profiles in which the thermal pressure force nearly balances the gravitational force at all radii (Chandrasekhar, 1939). The central parts of the core thus have much higher densities than the outer parts.

Observations of the star-forming regions indicate that higher mass cores have a more complicated structure than in the simple model described before (Myers and Fuller, 1992; Jijina et al., 1999). For sufficiently large size scales (> 3 light years) and low densities (< 10^4 atoms cm^{-3}), these observations suggest that the cores have substantial turbulent motions. The random velocities increase with the increasing density of the gas according a power-law relationship, which means that the velocities increase faster than the density. If these random velocities are interpreted as a transport speed, like the sound speed in air, then a turbulent component to the pressure can be derived (Lizano and Shu, 1989). In other words, the total pressure includes both a thermal and a nonthermal component, and both must be considered for high-mass cores, which provide the birth sites for higher mass stars.

Protostellar Collapse

During the protostellar phase, the star-like object at the center of the collapse flow is actively gaining mass and growing larger. For low-mass protostars, that is, forming stars with masses comparable to that of the sun, the radiation field is not strong enough to affect the infalling envelope, and the dynamic collapse is thus decoupled from the radiation. Molecular cloud cores tend to collapse from inside out (Shu, 1977); in other words, the central part of the core collapses first and successive outer layers follow. The collapse scenario naturally produces a core/envelope structure, where a pressure-supported object (the forming star itself) forms at the center and is surrounded by an infalling envelope of dust and gas. This inside-out collapse progresses as an expansion wave propagates outward at the sound speed. Outside the location of the expansion wave, the cloud remains nearly static and has no information that collapse is taking place in the interior. Inside the expansion wave radius, the material falls inward and approaches free-fall velocities.

The density distribution of this infalling envelope of dust and gas is nearly spherical outside a centrifugal radius, which we denote as R_C. The centrifugal radius provides an important length scale in the protostellar envelope: it defines the outer radius of the forming circumstellar disk, which later provides the initial conditions for planet formation. Its value is determined by the position where the infalling material with the highest **specific angular momentum** encounters a centrifugal barrier due to the conservation of angular momentum (Cassen and Moosman, 1981; Terebey *et al.*, 1984). Inside R_C, the collapse flow is highly nonspherical as particles spiral inward on trajectories controlled by gravity, not pressure. In the region immediately surrounding the star, the high temperatures evaporate the dust grains, so all the material is gaseous (Stahler *et al.*, 1980).

This collapse scenario contains no single, well-defined mass scale. Instead, the collapse flow feeds material onto the central star and disk at a well-defined mass infall rate. In the simplest case of an **isothermal** cloud core (i.e., with uniformly constant gas temperature), the mass infall rate is a constant in time and depends only on the initial temperature of the molecular cloud core. When additional sources of pressure are taken into account, the core structure is more complicated, and the mass infall rate generally increases with time. Typical mass infall rates lie in the range of 10^{-6}–10^{-4} $M_\odot$ yr^{-1}, so that the time required to make a star is a fraction of 1 million years. In fact, the larger mass infall rates tend to produce stars of higher mass, so that the star-formation time scale does not vary substantially with stellar mass. In addition, this formation time is extremely small compared with the lifetime of a Sun-like star and the age of the Universe (where both of these time scales are of order 10 billion years).

Protostellar Radiation

The radiation emitted by the forming stars (protostars) is important because it allows for a test of the underlying theory and for a means of identifying the protostellar candidates. The radiation field for protostellar objects can be divided into three separate components: the direct radiation field from the star, the direct radiation field from dust in the disk, and the diffuse radiation field emitted by the infalling dust envelope (see Figure 5.4). Both the star and disk can actively generate energy and thus emit radiation. Like our Sun, the young star emits most of its radiation at optically visible wavelengths. The disk also radiates at optical wavelengths, although it emits a substantial amount of energy in the infrared. However, much of the star/disk radiation is highly attenuated by the infalling envelope — most of the luminosity is absorbed by the dust grains in the envelope and then reradiated at longer (far-infrared) wavelengths that can escape more easily from the region. As result, the spectrum of radiation that we actually see is determined mostly by the properties of the infalling envelope and is largely independent of the spectrum of the initial star/disk system.

The protostellar luminosity has several different contributions, although the ultimate source of energy is gravity. At this early stage of stellar evolution, nuclear fusion of hydrogen is not yet taking place in the star. As material falls toward the central star/disk system, gravitational potential energy is converted to kinetic energy, which is then converted

Figure 5.4. A star in the process of formation. This artist's conception corresponds to the third stage (c) from Figure 5.2, where the central star is surrounded by a rotating disk of material and an outer envelope of dust and gas. In addition, the star generates a strongly focused wind or jet that propagates outward along the rotational poles — the vertical axis — of the system. (Image Credit: NASA/JPL-Caltech/R. Hurt (SSC)).

into radiation in several different ways. The infalling material produces shock waves on the surfaces of both the star and disk and thereby dissipates energy as heat. Additional energy is dissipated as the infalling interstellar material becomes adjusted to the stellar and disk conditions. In particular, the infalling material does not have the same rotation speeds as the material already in the star and disk; the newly added material must release energy as it adjusts to the local conditions. In addition, disk accretion produces a substantial amount of luminosity; in this process, the material in the circumstellar disk dissipates energy and transfers angular momentum outward. Some type of viscosity, which acts like friction on fluids, converts the rotational motion of the disk material into heat. The main uncertainty in this picture is the mechanism that enables the disk material to accrete onto the star, specifically, the source of the viscosity. However, disk-stability considerations place fairly tight constraints on the allowed disk accretion activity. The total luminosity of the system is thus reasonably well determined and is a substantial fraction of the total available luminosity, which can be written in the form

$$L_0 = \frac{GM_*\dot{M}}{R_*},$$

where R_* is the stellar radius (which is typically a few times the radius of the Sun) and M_* is the stellar mass (see Stahler et al., 1980; Larson, 1969). The total available luminosity L_0 given by the above equation represents material falling through a gravitational potential well (of depth GM_*/R_*) at the mass infall rate $\dot{M}$. The actual luminosity of the system is generally less than this maximum value, because some material does not fall all the way to the stellar surface, but remains in the orbit about the star at larger radii. As a result, some energy is stored in the form of rotational kinetic energy and gravitational potential energy.

Another interesting complication is that the disk accretion rate is not constant in time. Although the disk gains mass from the surrounding, infalling envelope at a steady rate, the disk tends to build up its mass and then dump it all on the star in bursts. As a result, the power output of forming stars is highly variable.

The diffused radiation field of the infalling dust envelope can be determined through a self-consistent radiative transfer calculation (Adams and Shu, 1986; Kenyon et al., 1993). Such a calculation keeps track of all photons (the particles that make up the radiation field) as they travel outward through the envelope and become absorbed by the dust grains and reradiated at longer wavelengths. The theoretical **spectral energy distributions** calculated from this protostellar model (Adams et al., 1987) are in reasonably good agreement with observed spectra of protostellar candidates (Myers et al., 1987). The spectral energy distributions show how much energy is emitted by the protostar at various wavelengths (or, equivalently, frequencies). These spectra generally have maxima at wavelengths of 60–100 μm, the far-infrared part of the spectrum. Almost all of the energy is emitted at wavelengths much longer than that of visible light; this property of the sources explains why protostellar candidates were identified only in the 1980s when infrared detectors became sufficiently developed. The spectra of sources in the bipolar outflow stage of evolution are also well described by protostellar infall models; this result suggests that both the infall and outflow are taking place simultaneously in such objects.

Because the density distribution of the protostellar envelope is known from the dynamic collapse solution and the temperature distribution is known from radiative transfer calculations, the spatial distribution of emission from protostars can also be calculated theoretically. Recent observations have produced spatial emission maps of protostars at millimeter and submillimeter wavelengths in the **continuum**. The measurements show that protostellar emission is spatially extended and that the observed spatial profiles (maps) of emission are roughly consistent with the current theory (Walker *et al.*, 1990; Butner *et al.*, 1991; Ladd *et al.*, 1991). In recent years, the Atacama Large Millimeter/submillimeter Array (ALMA) located in the Andes mountains in northern Chile, has made additional emission maps of protostellar objects with high resolution and sensitivity. A beautiful example, the ALMA map of the protostar HL Tau, with its surrounding disk, is shown in Figure 5.5.

Astronomers have traditionally studied and described stellar evolution by making use of the Hertzsprung–Russell (H–R) diagram, which plots the luminosity of a star on the vertical axis and the surface temperature of the star on the horizontal axis. These two physical variables (luminosity and surface temperature) are used because they adequately characterize a given star and because they can be determined directly from observations. Ordinary hydrogen-burning stars live on a well-defined locus in this diagram known as the **main sequence**; different points along the main sequence correspond to stars of different masses. Once stars have used up the hydrogen in their cores, they evolve into new stellar configurations; as they evolve, they follow well-defined paths (usually called tracks) in the H–R diagram. Thus, the H–R diagram provides a useful tool with which to study stellar evolution. However, protostars cannot be placed in the H–R diagram because they have no well-defined surface temperature. Protostars have extended atmospheres (their infalling envelopes) and hence have spectral energy distributions that are made up of emission from many different regions of different temperatures. As a result, another approach must be invoked to diagrammatically represent the early phase of stellar evolution (see Young and Evans, 2005).

The Protostellar to Stellar Transition

The transition from an embedded protostellar object to an optically revealed young star is still not completely understood. One piece of the puzzle is provided by the powerful stellar winds and outflows that are

The Origin of Stars and Planets

Figure 5.5. Image of a circumstellar disk surrounding the newly formed star HL Tau (image taken with the Atacama Large Millimeter Array; ALMA Partnership Brogan *et al.*, 2015).

observed in conjunction with the later stages of protostellar evolution. The presence of such winds (and related outflow phenomena) was first established observationally (Lada, 1985). The presently favored theoretical description of these phenomena is a centrifugally driven magnetic wind model (see Figure 5.6), although some controversy exists as to whether the wind originates near the star (Shu *et al.*, 1988; Shu *et al.*, 1994), or at larger disk radii (Königl and Pudritz, 2000).

This wind mechanism (illustrated by Figure 5.6) can be described roughly as follows: The central star generates a relatively strong magnetic field that rotates with the star. The field strength must be about 1000 Gauss, which is large compared with the value of about 1 Gauss for the Sun today (but we note that young stars are observed to have such strong fields). Some of the magnetic field lines are open: One end of the field line is anchored to the stellar surface; however, the field line does not attach back onto the star, but rather continues out to spatial infinity. Parcels of ionized gas can then flow along these field lines (but not across the field lines). In this setting, parcels of gas travel along the magnetic

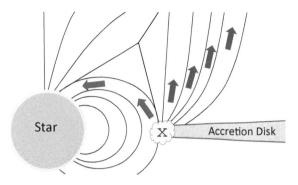

Figure 5.6. Schematic diagram for model of protostellar outflows (adapted from Shang, 2007). In this model, material flows inward through the accretion disk (from the right). At the X-region, the inner edge of the disk, some of the incoming material climbs onto magnetic field lines that connect to the star and then flows to the stellar surface, as depicted by the red arrows. The remaining material flows onto the open magnetic fields (as marked by the purple arrows) and gets flung outward at high speed.

field lines in much the same way that beads can travel along a wire. In this case, however, the magnetic field lines are rotating so that the parcels of gas are flung outward by the centrifugal force. These outwardly moving parcels of gas become the wind.

When the young pre-main-sequence stars first appear optically visible, so that they can be placed in the H–R diagram, they seem to appear along a well-defined locus, denoted as the stellar birthline (Stahler, 1983; Palla and Stahler, 1990). The location of this birthline corresponds to stellar configurations that are capable of burning **deuterium** (Shu, 1985; Stahler, 1988). Deuterium is the most easily fused of all available nuclei; deuterium-burning in stars can take place at a central stellar temperature T_c of approximately 10^6 K, whereas a much higher temperature, $T_c \approx 10^7$ K, is required for the fusion of ordinary hydrogen. In any case, stars appear on the H–R diagram as visible objects with just the right properties to burn deuterium. One possible explanation for this finding is that deuterium-burning produces stellar **convection**; in other words, the heat generated by deuterium-burning is carried toward the surface of the star by motions in the stellar fluid itself. These convective motions, in conjunction with differential rotation in the star, can generate and amplify magnetic fields to the large strengths (100–1000 Gauss) observed in young stars. Moreover, magnetic field strengths of this magnitude are required to drive the centrifugally driven winds described above.

The Stellar Initial Mass Function

The IMF is the distribution of masses of a given stellar population at the moment of birth. It is perhaps the most fundamental output of the star-formation process. A detailed knowledge of the IMF is required to understand galaxy formation, the chemical evolution of galaxies, the evolution of the interstellar medium, and other important astronomical issues (see Chapters 2 and 3). Unfortunately, at the present time, we remain unable to calculate the IMF from *a priori* considerations. We can, however, empirically can determine the IMF shape in our galaxy today. As a first approximation (Salpeter, 1955), the number of stars born with mass m in the mass range m to $m + dm$ is given by the simple power-law relation

$$\frac{df}{dm} \alpha\, m^{-\gamma},$$

where the index $\gamma = 2.35$ for stars in the mass range $m = 0.4$ to $10\ M_\odot$. Many more stars are born with small masses than with larger masses. To make this point precise, stars with masses 10 times that of our Sun are about 220 times less numerous than stars with the mass of the Sun. Subsequent works (Miller and Scalo, 1979; Scalo, 1986) showed that the mass distribution is somewhat more complicated than a simple power law. The distribution has a maximum value near $m = 0.25\ M_\odot$ (Kroupa, 2001; Chabrier, 2003) and decreases for even lower masses. Because of this turnover, fewer stars have masses near the minimum mass of $m = 0.08\ M_\odot$ than the peak value $m = 0.25\ M_\odot$.

Stars can only exist in a finite range of masses. Stellar objects with masses less than about $m = 0.08\ M_\odot$ cannot produce central temperatures hot enough for the fusion of protons (the most common form of hydrogen) to take place. Objects with masses less than this limit are known as brown dwarfs. On the other end of the possible mass range, stars with masses greater than about $m = 100\ M_\odot$ cannot exist because they are unstable (see Phillips, 1994, for a textbook description). As a result, stellar masses are confined to the particular range

$$0.08\ M_\odot \leq m \leq 100\ M_\odot.$$

Although the upper limit is approximate, this mass range is much smaller than the conceivable mass range. Stars form within galaxies that have masses of about $10^{12}\ M_\odot$ and stars are made up of hydrogen atoms that can have masses of about 10^{-24} grams, or approximately 10^{-57} $M_\odot$. Thus, galaxies could build objects anywhere in the mass range from 10^{-57} to 10^{12} $M_\odot$, a factor of 10^{69} in mass scale! Yet stars live in the mass range given above, which allows stellar masses to differ from each other by a factor of only 1,000.

Although we do not have a complete theory that determines the stellar IMF, we can conceptually divide the process into two basic subproblems: the spectrum of initial conditions produced by the interstellar medium, and the transformation between a given set of initial conditions and the properties of the final (formed) star.

To understand the spectrum of initial conditions, we must understand the physical processes that produce the distribution of properties of molecular cloud cores, which provide the sites of individual star formation events. The mass distribution of these cores is roughly similar in shape to that of the stellar IMF (Rathborne et al., 2009), but the mass scale is larger by a factor of 3 or 4. This observational finding provides a promising start to specifying the distribution of initial conditions for star formation. To move forward, we need to determine the distributions of other relevant core properties, including rotation rates, density profiles, turbulence levels, and magnetic field configurations. In addition to measuring the distribution of these properties for molecular cloud cores, we must also understand the physical processes that determine these distributions. More work on the formation of substructure within molecular clouds — and the life cycles of the clouds themselves — is needed before we can begin to predict the distribution of core properties starting from basic physical principles.

We also need to understand the transformation between the initial conditions specified by molecular cloud cores and the final masses of the stars they produce. On the scale of the molecular clouds, only a few percent of the total mass is processed into stars during a free-fall time. On the scale of the cores themselves, a sizable majority of the available mass does not end up in the stars that are produced. Most of the mass remains behind.

The processes that prevent all of the mass in a core from collapsing into the stars they form remain under study. Because most of the core mass does not become part of the star, stellar masses are *not* determined simply by having the mass supply run out. Conservation of angular momentum plays an important role, as most of the mass falls onto a rotationally supported disk, rather than directly onto the stellar surface, so that disk physics provides one bottleneck in the process. Some combination of thermal pressure, turbulence, and magnetic fields provides another impediment. Finally, as outlined briefly before, young stars produce powerful outflows that eventually have enough momentum to separate newly formed stars from their immediate environment.

Summary of the Star-Formation Paradigm

Observations of protostellar candidates are in good agreement with theoretical expectations. The spectral energy distributions calculated from the protostellar theory agree with the observed spectra for objects in both the pure infall and the bipolar outflow phases of evolution; hence, infall is probably still taking place in the bipolar phase. The emission from protostellar candidates is observed to be extended in a manner roughly consistent with theoretical expectations. In addition, the spectral energy distributions of T Tauri stars with infrared excesses can be understood as young stars surrounded by circumstellar disks. Some systems have passive disks that reprocess stellar radiation but have no intrinsic luminosity; other systems require appreciable intrinsic disk luminosity. For some sources, estimates for the disk masses (M_D = 0.01 to 0.2 $M_\odot$) and the disk radii ($R_D \approx$ 100 AU) can be obtained. Recall that 1 AU is the distance from the Earth to the Sun, by definition, and that the size of our Solar System, given by the radius of Neptune's orbit, is about 30 AU.

The existing theory of star formation is not yet complete. In particular, we still must understand more about the disk accretion mechanism as well as the outflow mechanism and the manner in which newly formed stars are separated from their birth sites. The current version of the theory works well for the formation of single stars of low mass, which includes stars like our Sun. However, many stars are members of binary systems (Abt, 1983) and stellar masses extend up to ~ 100 times that of the Sun (Phillips, 1994). The extension of the theory to include the formation of binary systems (see Duchêne and Kraus, 2013 for the current observational picture) and stars of higher mass is being studied (Zinnecker and Yorke, 2007).

Circumstellar Disks and Pre-Main-Sequence Stars

During the protostellar phase of evolution, disk accretion must occur because most infalling material falls onto the disk rather than directly onto the star. In the absence of some mechanism to transfer mass from the disk to the star, the masses of the forming stars would be unrealistically small, the luminosity would be much smaller than that of observed sources, and the disk would become gravitationally unstable. During the T Tauri phase of evolution, disk accretion must also occur to produce the observed intrinsic disk luminosities and to account for the fact that disks seem to disappear on relatively short time scales (Haisch *et al.*, 2001). Although the disk accretion mechanism is not fully understood, two important ingredients that drive disk evolution are gravitational instabilities and viscosity.

Any given star/disk system tends to evolve toward a configuration of lower total energy. This trend holds true for any kind of dissipative mechanism. On the other hand, the system must also conserve its total angular momentum. The lowest energy state accessible to the system is to have most of the mass in the central star — so that the material resides as deep into its gravitational potential well as possible — with a small mass in a large-radius orbit to carry all the angular momentum. We expect any star/disk system to evolve toward, but not necessarily reach, this minimum energy configuration. This state-of-affairs is very nearly met within our Solar System — almost all of the mass resides in the Sun itself, and almost of all the angular momentum is carried by the orbital motion of the giant planets (which orbit at large radii). All planetary systems are expected to evolve in a similar manner. The key question is how the energy dissipation and redistribution of angular momentum takes place.

Radiation from Circumstellar Disks

Ordinary stars have spectral energy distributions with the basic form of a blackbody. As mentioned earlier, however, young stars are often observed to have an additional infrared component to their spectra (Rydgren and Zak, 1987; Rucinski, 1985; Appenzeller and Mundt, 1989), and circumstellar disks provide the most likely explanation for this infrared excess. These disks can either be active or passive. Active disks generate their own energy through the process of disk accretion; passive disks have no intrinsic energy source and only absorb and reradiate stellar photons. In either case, the disk radiates photons at infrared wavelengths and produces the observed excess emission.

In systems with passive disks, all of the luminosity is generated by the star itself. The disk is generally spatially thin and optically thick, so that the disk absorbs all the light that is incident upon it. When the inner radius of the disk extends down to the stellar surface and the outer disk radius is large compared with that of the stellar radius, and when the disk is perfectly flat, it intercepts and reradiates 25% of the stellar luminosity. In practice, disks are not flat, but rather bend upward and intercept even larger quantities of stellar light. Because of this intercepted light, the disk radiates as if it had its own effective luminosity (one-fourth to one-half that of the star) and adds excess infrared radiation to the spectral energy distribution of the system. The passive disk models produce the correct infrared excess for some observed T Tauri systems (Adams *et al.*, 1987; Kenyon and Hartmann, 1987; Chiang and Goldreich, 1997).

The class of star/disk systems with active disks have appreciable intrinsic disk luminosity in addition to radiation intercepted and reprocessed from the star. These systems are thus actively generating energy through disk accretion. Through fitting their spectral energy distributions, we can estimate the basic physical properties of active disk systems (Kenyon and Hartmann, 1987; Adams *et al.*, 1988; Beckwith *et al.*, 1990; Adams *et al.*, 1990; Calvet *et al.*, 1994; D'Alessio *et al.*, 1998). In extreme cases, the energy generated by the disk is comparable to that generated by the star itself.

We can also obtain estimates for the minimum outer radii of these disks, approximately 100 AU, which makes them comparable in size to our solar system. Estimates for the disk masses (M_D) can be obtained by measuring the radiation spectrum at millimeter wavelengths where the amount of radiation observed is directly proportional to M_D. Currently available estimates show an upper limit for disk masses of approximately $M_D = 0.10$ $M_\odot$ (starting with Adams *et al.*, 1990), and continuing down to much smaller values (Beckwith *et al.*, 1990). However, these measured disk masses only include particles with sizes up to about 1 centimeter. Larger particles are essentially invisible to the radio telescopes used for these measurements even though these larger particles and even larger objects are believed to contain most of the disk masses over much of the disk lifetimes. Star/disk systems live in the active phase for somewhat less than 1 million years, and retain their gas for several million years thereafter (Haisch *et al.*, 2001). The estimated disk properties (radial sizes, masses, and angular momenta) are in good agreement with the disk properties predicted by the protostellar theory. Moreover, these disk properties are consistent with those required to form a solar system such as our own.

Gravitational Instabilities

Gravitational instabilities in circumstellar disks provide one important mechanism that can lead to energy dissipation and transfer of angular momentum, which then leads to accretion of material from the disk and onto the central star. In addition, if the instabilities grow to sufficiently large amplitudes, they can also lead to the formation of giant planets or even larger companions to the star.

All disk systems tend to produce spiral patterns, and those associated with star formation are no exception. Well-known examples of this type of behavior are spiral galaxies and the spiral patterns observed in the rings of Saturn. The growth and behavior of spiral instabilities are mainly determined by three elements: gravity, pressure, and differential rotation.

Star/disk systems often support spiral patterns with only one arm, in contrast to galaxies where the patterns generally have two arms. Because the central star contains most of the mass, and dominates the gravitational potential, parcels of gas in the disk travel on simple orbits that are analogous to the planets orbiting the Sun in our Solar System. In this case, the particle orbits are simple closed ellipses. In the absence of other forces, the orbits would not change with time. Moreover, if the orbits are lined up to form a spiral pattern, they will stay in that configuration. However, the disk material is subject to other forces, such as pressure, which tend to spread out the spiral pattern. On the other hand, the self-gravity of the disk acts to hold the spiral together and sustain the pattern. As a result, spiral patterns arise naturally in these circumstellar disk systems. Keep in mind that the spirals are wave patterns and are not material arms (which would wrap up with time due to the differential rotation of the disk). In other words, the same molecules do not stay in the spiral arms as they orbit around the star, just as the same water molecules do not stay in a tsunami as it propagates across the ocean.

Patterns with a single spiral arm are especially interesting in the context of forming star/disk systems because the center of mass of the spiral shape does not coincide with the star. This effect gives rise to a new forcing mechanism (a new way to drive wave motions through the disk), which plays an important role in the amplification of these types of spiral modes (Adams *et al.*, 1989; Shu *et al.*, 1990). The new forcing mechanism is essential for the growth and maintenance of spiral modes with one spiral arm. In addition, the asymmetry causes the star to move outward from the center of mass as the disk transfers angular momentum to the stellar orbit. This mechanism could thus help in the formation of binary companions and/or giant planets. An example of a spiral pattern in a cirumstellar disk around a protostar is shown in Figure 5.7.

The Origin of Stars and Planets

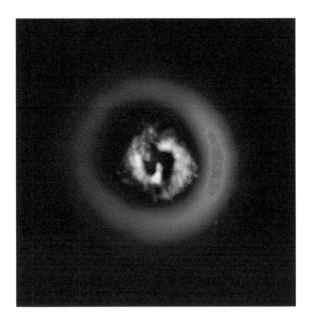

Figure 5.7. A high resolution map made by the millimeter-wave interferrometer ALMA, of the gas disk around the star AB Aurigae. The outer disk of cold dust is visible in red. The inner region has warm molecules including CO, whose emission is shown in blue. This warmer denser gas shows a spiral pattern as described in the text. *Source*: National Radio Astronomy Observatory.

From the combined results of many recent studies, we now have a basic understanding of the physics of self-gravitating instabilities in disks. In particular, we can determine the dependence of the growth rates on the underlying physical parameters of the star/disk system. Disks become more stable as the temperature is raised (because of the increase in pressure support). On the other hand, disks tend to become more unstable as the total disk mass is increased (because of the increase in gravitational force).

The growth rates of the instabilities are largest when the disk and star have equal masses ($M_D = M_*$) and decrease rapidly with decreasing relative disk mass. In the optimal case, $M_D = M_*$, the instabilities would grow on a time scale comparable to the orbital time at the outer disk edge. In other words, the modes would grow on nearly the dynamical time scale. This time scale is typically a few thousand years, which is much shorter than the evolutionary time scale (millions of years) of the disks.

Thus, these gravitational instabilities can grow fast enough to significantly affect disk evolution.

For the particular case of one-armed spiral modes, a minimum mass threshold exists for strong amplification of the instabilities. This threshold implies a critical value of the relative disk mass, the maximum value of the ratio M_D/M_*, that is stable to these disturbances. For the simplest kind of disk, where gas orbits are entirely determined by the balance between gravity and centrifugal acceleration, the critical ratio has the value M_D/M_* ~ 1/3. In other words, when the disk mass is greater than about one-third the stellar mass, gravitational instabilities grow strongly and the disk is highly unstable. When the disk mass becomes less than about one-third the stellar mass, instabilities can still grow, but at a much slower rate.

Realistic Simulations of Star/Disk Systems

To study the advanced behavior of gravitational instabilities, after they have grown large density inhomogeneities, one must perform numerical simulations. Results have been obtained from several different hydrodynamic simulations of gravitational instabilities in simple disks (Laughlin and Bodenheimer, 1994; Boley *et al.*, 2006). In some simulations, computer-generated "particles" are used to represent parcels of gas. All the simulations assume a uniform gas temperature; from a physical point of view, this implicitly assumes that the disk is able to efficiently radiate all energy dissipated during the evolution. Results show that gravitational instabilities can grow strongly in these systems and can produce well-defined spiral patterns, much like those found in spiral galaxies. In particular, the growth rates are comparable to the orbital time scale of the outer disk edge and are thus in agreement with the stability calculations described earlier. For disks that are not too far from the condition of stability, spiral instabilities with different numbers of spiral arms (m = 1,2,3,4 and higher) develop. As stability is increased (i.e., as the temperature is raised or the mass ratio M_D/M_* is decreased), the relative strength of the one-armed disturbance increases, although the growth rates of all modes decrease as expected.

A spiral arm can collapse to form a clump of gravitationally bound gas when the instability in the disk becomes sufficiently strong. These collapsed clumps typically have masses of about 0.01 M_* and travel on elliptical orbits. The possibility that the clumps survive to form either giant planets or binary companions is especially interesting. If the clumps eventually form binary companions, they must gain mass from the disk faster than does the original star. On the other hand, if the clumps eventually form giant planets, the resulting bodies would have the same composition as their host stars. This result stands in contrast to the giant planets in our solar system, which contain large rocky cores and are heavily enriched in heavy elements relative to the Sun. Although more work is necessary to understand the long-term fate of such clumps, the most likely result is the production of brown dwarfs. These objects are more massive than planets but less massive than stars, and are sometimes found in distant orbits around young stars.

However, these clumps are not always produced from all types of self-gravitating instabilities (Laughlin and Bodenheimer, 1994; Woodward et al., 1994). As a general trend, to form a gravitationally bound clump in a circumstellar disk, the underlying perturbation must be at least 3 or 4 times denser than the average surroundings. For perturbations with smaller density contrasts, the clumps do not form. Instead, the effects of angular momentum transfer dominate, and the spiral modes drive an accretion flow through the disk. Three time-frames from a recent computer simulation of these self-gravitating instabilities in a circumstellar disk are shown in Figure 5.8.

Viscous Evolution of Circumstellar Disks

In the later stages of disk evolution, the disk mass must eventually become small enough that gravitational instabilities turn off, or at least become substantially less important. For the case of intermediate mass disks (i.e., when the disk mass is too small for gravitational effects to dominate, but large enough for planet formation and other disk processes to take place), we expect circumstellar disks to evolve through the action of viscosity. Such systems are generally known as viscous accretion disks. Any source of fluid viscosity (which is essentially a frictional force) creates both energy dissipation and transfer of angular momentum; these are precisely the two physical effects that must occur for disk accretion to take place.

The most important issue in this picture is the source of the fluid viscosity. Ordinary molecular viscosity is always present and arises from frictional forces between molecules; however, this viscosity is much too small to be of astrophysical importance. Thus, a source of anomalous viscosity must somehow be generated. Many different mechanisms have been proposed to generate turbulence in circumstellar disks; turbulence,

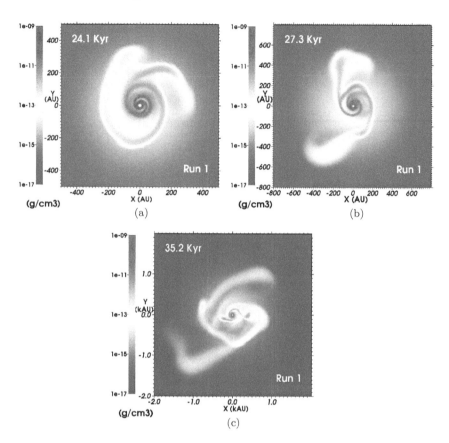

Figure 5.8. Snapshots from a computer simulation of the evolution of a gas disk around a young star. The gas density in the central thousand astronomical units is shown with color coding. By 24,000 years (upper left panel), a strong spiral arm has formed. It has the highest density, shown by dark red. By 35,000 years (bottom panel), this over-density has broken up into several individual clumps, which could collapse to form giant planets or perhaps brown dwarfs. *Source*: Monthly Notices of the Royal Astronomical Society.

in turn, leads to small-scale motions that allow for the dissipation of energy and, hence, an effective viscosity.

The leading candidate for generating turbulence is through an instability that results from the interplay between magnetic fields and the differential rotation in the disk (Balbus and Hawley, 1998). The mechanism is called the magneto-rotational instability (MRI). In this scenario, the magnetic field, which is relatively weak, provides a means of transferring energy from the orbital motion in the disk into small-scale motions

(Balbus and Hawley, 1991). This process can occur whenever the energy density of the magnetic field is less than the thermal energy density. The resulting instability leads to turbulence in the disk and, in turn, provides a means of transporting angular momentum.

The evolution of viscous accretion disks is governed by the laws of fluid dynamics. In the limit that viscous forces drive the evolution of the disk, the behavior is described by a time-dependent diffusion equation (Liist, 1952; Lynden-Bell and Pringle, 1974; Lin and Papaloizou, 1985). In other words, disk evolution occurs through a viscous diffusion process. As in any diffusion process, the net result is for the system (here, the disk) to spread out. In this case, the inner parts of the disk move farther inward and some disk material is accreted onto the star in the center. On the other hand, the outer parts of the disk gain angular momentum and spread farther outward. As this process continues, the disk mass becomes smaller relative to the star, the surface density of the disk decreases, and the outer radius of the disk increases. Thus, this process makes a star/disk system evolve toward the current state of our own solar system, with most of the mass in the central star and most of the angular momentum carried by objects at large radii (the planets). The evolutionary time scale depends on the size of the viscosity.

Summary of Disk Processes

Several different processes can lead to angular momentum transport and hence accretion in circumstellar disks. Self-gravitating instabilities can grow on dynamical time scales (as short as a few thousand years, the orbital time scale at the outer disk edge), which is much shorter than the evolutionary time scale of these systems (millions of years). Computer simulations show that these instabilities can grow well into the nonlinear regime, so that the density of the growing spiral arms becomes much larger than the initial density in the disk. For disks where the perturbations become sufficiently robust, small clumps of gravitationally bound gas can form out of the disk. If these clumps survive, they can become either planets, brown dwarfs, or even stellar companions to the original star. The most natural mass-scale for such bodies is intermediate between the mass of Jupiter and a star, so that brown dwarf companions are a likely outcome. Although such bodies have been observed (e.g., Marois *et al.*, 2008, found objects that are either large planets or small brown dwarfs), they are rare, so that most disks fail to produce lasting clumps.

For disks with less successful perturbations that halt their growth at lower amplitudes, instabilities drive angular momentum transport and disk accretion. As a result, gravitational instabilities play a role in the process of disk accretion and sometimes the formation of companions (e.g., brown dwarfs or large planets).

Gravitational instabilities are active early in disk evolution, while the disk retains a substantial mass. At sufficiently late times, when the disk mass is smaller, gravitational instabilities shut off and some type of viscous accretion takes over. The presence of any type of viscosity leads to viscous diffusion of the disk. Current thinking suggests that magnetic instabilities lead to turbulence, which produces the requisite viscosity (Balbus and Hawley, 1991, 1998). Under this action, the disk spreads out and produces an accretion flow from the disk onto the star. At sufficiently long times, the star/disk system is driven to a final state much like that of our solar system — most of the mass ends up in the central star and most of the angular momentum resides in the orbital motion of planets at large radii.

Planet Formation

The hypothesis that planets form within the circumstellar disks that surround young stellar objects — known as the nebular hypothesis — was first stated more than two centuries ago (Kant, 1755; Laplace, 1796) and continues to be the most viable scenario for planet formation. As outlined above, circumstellar disks form naturally as part of the star-formation process. Moreover, these disks have the proper masses and physical sizes to provide the initial conditions for planet formation.

We can identify two conceptually different ways for planets to form within a circumstellar disk. The first of these processes, accumulation of **planetesimals**, assumes that planets form by the gradual accretion of small, rock-like bodies. In this case, the planets form "from the bottom-up." Alternatively, planets could also form through a gravitational instability in the disk. In this case, the circumstellar disk becomes gravitationally unstable and breaks into secondary bodies that become planets. In this second scenario, the planets form "from the top-down." Although both planet-formation scenarios have some difficulties, the accretion of planetesimals is generally thought to produce the majority of planets.

Our solar system contains two different types of planets. The inner four, Mercury, Venus, Earth, and Mars, are known as the terrestrial planets. They are composed primarily of heavy elements (i.e., elements heavier than hydrogen and helium, which make up 99% of the mass of the Sun). The next four planets, Jupiter, Saturn, Uranus, and Neptune, are known as the giant planets or the Jovian planets. They are much more massive than the terrestrial planets and contain substantial amounts of hydrogen and helium. They do not, however, contain as much hydrogen and helium as the Sun; they are highly enriched in heavy elements relative to the Sun. The giant planets also differ from stars (and the Sun) in that they contain solid rocky cores in their centers. The elemental abundances of the different planets place important constraints on how they must have formed. In addition, our solar system contains many other bodies, including asteroids, Pluto and other dwarf planets, smaller Kuiper Belt objects orbiting beyond Neptune, and comets in even more distant orbits.

The circumstellar disk from which the planets formed must have contained a substantial amount of mass. Clearly, the mass of this disk must have been at least as large as the total mass of all our planets and other solar system bodies. However, because these bodies contain relatively more heavy elements than the Sun itself, the initial total disk mass must have been at least as large as the total mass of all solar system bodies when augmented to solar abundances, that is, when the missing hydrogen and helium are added back into the total. The augmented disk would have had a mass of about 0.01 $M_\odot$, a benchmark scale known as the minimum mass solar nebula.

In the past two decades, astronomers have detected thousands of new worlds in orbit about other stars (Batalha *et al.*, 2013). These discoveries show that planetary systems can be built with a diverse set of different architectures. The first planets found in orbit around main-sequence stars (Mayor and Queloz, 1995) included hot Jupiters — planets with masses comparable to our Jupiter, but with orbits of only a few days, thereby placing them extremely close to their parental stars. Other planetary systems contain Jovian planets with longer periods, but with highly elongated (eccentric) orbits. Still other planetary systems show tightly packed collections of planets with intermediate masses. These planets are larger than Earth and smaller than Neptune; although they are not represented within our own solar system, these super Earths (as they are often called) may be the most common outcome of the planet-formation process. Astronomers have finally discovered a planet with mass as small as Earth and with an orbit that places it inside the habitable zone where liquid water could exist

Origin and Evolution of the Universe

Figure 5.9. Artist's conception of Kepler-186f, the first validated Earth-size planet to orbit a distant star in the habitable zone — the range of distance from a star where liquid water could exist on the planetary surface. The discovery of Kepler-186f (Quintana et al., 2014) demonstrates that Earth-sized planets exist in the habitable zones of other stars. In addition to Kepler-186f, thousands of planets have now been discovered in orbit around other stars. (Image Credit: NASA Ames/JPL-Caltech/T. Pyle).

(Quintana et al., 2014) (see Figure 5.9), although the host star is smaller than our Sun. These discoveries demonstrate that planetary systems can take many different forms. Although a comprehensive discussion of exoplanets is beyond the scope of this chapter, the basic constituents of planetary systems are the planets themselves. As outlined in the following, we can understand their formation in terms of two physical mechanisms.

Formation of Planets by Accumulation

Let us first consider planet formation through the accumulation of small solid bodies. In disks that form planets, most of the heavy elements are found initially in the form of dust grains. In the interstellar medium (before gas and dust become incorporated into star and disk), dust grains are very small, with typical radii about 10^{-5} centimeters. In the denser environment of a circumstellar disk, the grains are expected to become somewhat larger, but remain mostly microscopic in size (small compared with 1 centimeter). The process of planet formation thus begins with tiny dust grains and somehow produces huge objects with characteristic sizes of about 10^9 centimeters.

Dust grains accumulate on a fairly short time scale into large rock-like bodies called planetesimals. In the absence of turbulence, the dust grains settle to the midplane of the disk because they do not feel the same pressure force as the gas. In this case, the resulting thin layer of dust becomes gravitationally unstable and breaks up into planetesimals. The planetesimals are basically large rocks with sizes about 1 kilometer, that is, about the size of an asteroid. In other words, the planetesimals are much larger than the original dust grains, but are still much smaller than a planet. The picture of planetesimal production is complicated by the presence of turbulence, which prevents the dust grains from settling all the way to the disk midplane and inhibits gravitational instability. In this case, sticking together of dust grains becomes important. However, the net result is similar in that large rocks of roughly asteroid size and mass can be produced quickly. The time scale for the production of planetesimals is about 10^4 years (Johansen *et al.*, 2007), which is short compared with the expected lifetime of the disk (roughly 10^7 years).

The next stage of planet formation involves the accumulation of planetesimals into the planets themselves. The terrestrial planets are composed mostly of the rocky material that makes up the planetesimals, so their formation can be understood entirely in terms of planetesimal accumulation. However, the giant planets also contain substantial amounts of the lighter gases. For the latter type of composition to arise, the planetesimals must accumulate into a massive rocky core that subsequently accretes gas from the circumstellar disk. Giant planets thus require an additional stage of formation.

The accumulation of planetesimals into planets — and the cores of giant gaseous planets — takes considerably longer than the initial buildup of the planetesimals. One relevant time scale for this process is the orbit time at the radial position of the forming planet. This time scale is 1 year at the radius of Earth's orbit, 1/4 year at the radius of Mercury's orbit, and 164 years at the radius of Neptune's orbit. As a result, the natural clock runs considerably slower in the outer solar system compared to the inner parts. Another important effect is that the density of the disk (and, hence, the number density of the planetesimals) decreases with radius. These properties combine to make the rate of planetesimal collisions much smaller in the outer solar system. As a result, it takes a long time for planetesimals to accumulate in the outer solar system. These same issues — slower orbits and lower densities in the disk — make it harder for large planets to form around small stars (Laughlin *et al.*, 2004).

As mentioned earlier, the time required for planetesimals to accumulate into planet-sized bodies is not completely understood. As a rough estimate, under the most favorable circumstances, the cores of the giant planets can be produced in about 10^6 years. Because the gas in the disk has a lifetime at least this long, the cores can accrete enough gaseous material to account for the observed compositions and frequency of giant planets. As a result, this scenario for planet formation basically works. However, not all disks will be able to produce giant planet cores during the time they retain their gaseous component. As a result, the formation of gaseous planets like Jupiter will sometimes fail. The result of such failure is a gas-poor planet like Neptune. Indeed, the inventory of extra-solar planets shows that Neptunes are more common than Jupiters (Winn and Fabrycky, 2015).

Formation of Planets by Gravitational Instability

Planets may also form through a gravitational instability in the disk. Gravitational condensation alone cannot account for the composition of planets with rocky cores — gravity is an equal opportunity force and does not distinguish between gas and rocky particles. As a result, any objects formed through the action of gravity alone are expected to have the same composition as their parental star, rather than being enriched in heavy elements. In principle, however, it remains possible for giant planets to form via gravitational condensation in the disk, provided that some other mechanism can produce the observed enrichment in heavy elements.

As discussed already, circumstellar disks can be unstable if their temperatures are low enough and their masses are large enough. Furthermore, some of the observed star/disk systems seem to live near this threshold for gravitational instability. The natural mass scale for a gravitationally bound object that forms in a disk can be shown to be roughly 0.01 $M_\odot$, about 10 times the mass of Jupiter, for typical star/disk systems. This mass is more than enough to form a giant planet, but too small to make a stellar companion. Moreover, the relevant time scale for gravitational condensation is comparable to the orbital time scales of the outer disk, thousands of years, much shorter than the accumulation time scale. As a result, the gravitational instability mechanism has both a time scale and a mass scale that are appropriate for the formation of brown dwarfs, and formation is most likely to occur in the outer realms of the disks, with orbital radii of about 100 AU (Boss, 1997; Rafikov, 2005).

The difficulty in this scenario of planet formation is that it naturally produces secondary bodies with stellar composition, rather than planets that are enriched in heavy elements. Moreover, our giant planets, and many others orbiting other stars, are inferred to have large rocky cores; the gravitational instability mechanism does not account for the production of such cores. Dust grains, which initially contain the heavy elements, naturally tend to settle toward the center of a self-gravitating protoplanet because the grains feel less force from pressure than the surrounding gas. However, the time scale for the dust-settling process is generally thought to be too long to account for the observed composition and structure of giant planets. Other mechanisms to provide dust separation have also been proposed (for example, dust settling in vortices), but core formation remains problematic.

Summary and Discussion

This chapter outlines the current theory of star formation in molecular clouds. We can conceptually divide the star-formation process into four stages (see Figure 5.2). In the first stage, molecular clouds produce small dense regions called molecular cloud cores. In the next stage, a core collapses to form a star/disk system that is deeply embedded in an infalling envelope of dust and gas. The star/disk system develops a powerful outflow (wind) in the next stage. Our current theory suggests that stars, in part, determine their own masses through the action of the powerful outflows. In other words, the outflows help separate a newly formed star/disk system from its molecular environment. In the fourth, final stage of star formation, the optically revealed star evolves into a configuration capable of sustaining the nuclear fusion of hydrogen. Also during this stage, the circumstellar disk transfers a substantial portion of its mass onto the star, produces planets, or both. The entire star-formation process takes place on a time scale of only a few million years.

This theory of star formation has many successes, especially for stars of low mass. In particular, the theory predicts both spectral energy distributions and spatial distributions of emission in good agreement with observations. The presence of circumstellar disks, which often must have substantial activity, can explain the spectral appearance of newly formed stars with infrared excesses. Direct observations of these disks have now been made. The estimated masses and radii of the circumstellar disks are consistent with those properties necessary to form planetary systems similar to our own.

In spite of its successes, the theory of star formation remains incomplete. One important unresolved issue is the mechanism that leads to accretion through the disks. We now understand the basic physics by which gravitational instabilities and viscosity can lead to disk accretion. For the former, however, we must understand the long-term evolution. For the latter, we must understand the source of the viscosity. In addition, the theory still must be extended to include the formation of high-mass stars and the formation of binary companions. We have just begun to calculate and predict the Initial Mass Function (IMF) for forming stars as a function of environmental parameters.

We began this chapter by noting that star formation and stellar evolution can be considered as a war between gravity and entropy. The battle takes place on many different size scales: in molecular clouds, in molecular cloud cores, in circumstellar disks, and within the stars themselves. The force of gravity must win at least a partial victory for stars and planets to form. However, the process of star formation ends with the production of a main sequence star, which represents a truce between the two warring parties. In these stars, the gravitational forces are exactly balanced by pressure forces generated through nuclear processes. The apparent state of peace is only temporary. Stars eventually run out of their nuclear fuel and must readjust their configurations; in other words, the war between gravity and entropy starts up again at the end of a star's life. The final readjustment can lead to violent explosions and the production of exotic stellar objects such as white dwarfs, neutron stars, and black holes (see Chapter 4).

References

Abt, H. A. 1983. Normal and abnormal binary frequencies. *Annual Review of Astronomy and Astrophysics* 21: 343–372.

Adams, F. C. 2010. The birth environment of the solar system. *Annual Review of Astronomy and Astrophysics* 48: 47–85.

Adams, F. C., Emerson, J. P., and Fuller, G. A. 1990. Submillimeter photometry and disk masses of T Tauri disk systems. *Astrophysical Journal* 357: 606–620.

Adams, F. C., and Fatuzzo, M. 1996. A theory of the initial mass function for star formation in molecular clouds. *Astrophysical Journal* 464: 256–271.

Adams, F. C., Lada, C. J., and Shu, F. H. 1987. Spectral evolution of young stellar objects. *Astrophysical Journal* 312: 788–806.

Adams, F. C., Lada, C. J., and Shu, F. H. 1988. The disks of T Tauri stars with flat infrared spectra. *Astrophysical Journal* 326: 865–883.

Adams, F. C., Ruden, S. P., and Shu, F. H. 1989. Eccentric gravitational instabilities in nearly Keplerian disks. *Astrophysical Journal* 347: 959–975.

Adams, F. C., and Shu, F. H. 1986. Infrared spectra of rotating protostars. *Astrophysical Journal* 308: 836–853.

ALMA Partnership, Brogan, C. L., Perez, L. M., Hunter, T. R., *et al.*, 2015. The 2014 ALMA long baseline campaign: First results from high angular resolution observations toward the HL Tau region. *Astrophysical Journal* 808: L3–L12.

Appenzeller, I., and Mundt, R. 1989. T Tauri stars. *Astronomy and Astrophysics Reviews* 1: 291–324.

Arons, J., and Max, C. 1975. Hydromagnetic waves in molecular clouds. *Astrophysical Journal* 196: L77–L82.

Balbus, S. A., and Hawley, J. F. 1991. A powerful local shear instability in weakly magnetized disks. I. Linear analysis. *Astrophysical Journal* 376: 214–222.

Balbus, S. A., and Hawley, J. F. 1998. Instability, turbulence, and enhanced transport in accretion disks. *Reviews of Modern Physics* 70: 1–53.

Batalha, N. M., Rowe, J. F., Bryson, S. T. *et al.*, 2013. Planetary candidates observed by Kepler. III. Analysis of the first 16 months of data. *Astrophysical Journal Supplement* 204: 24–44.

Beckwith, S., Sargent, A. I., Chini, R., and Gusten, R. 1990. A survey for circumstellar disks around young stars. *Astronomical Journal* 99: 924–945.

Bergin, E. A., and Tafalla, M. 2007. Cold dark clouds: The initial conditions for star formation. *Annual Review of Astronomy and Astrophysics* 45: 339–396.

Blitz, L. 1993. Giant molecular clouds. In Levy, E., and Mathews, M. S., (Eds.), *Protostars and Planets III* (Tucson: University of Arizona Press), pp. 125–162.

Boley, A. C., Mejia, A. C., Durisen, R. H., Cai, K., Pickett, M. K., and D'Alessio, P. 2006. The thermal regulation of gravitational instabilities in protoplanetary disks. III. Simulations with radiative cooling and realistic opacities. *Astrophysical Journal* 651: 517–534.

Boss, A. 1997. Giant planet formation by gravitational instability. *Science* 276: 1836–1839.

Butner, H. M., Evans, N. J., Lester, D. F., Levreault, R. M., and Strom, S. E. 1991. Testing models of low-mass star formation: High resolution far-infrared observations of L1551 IRS5. *Astrophysical Journal* 376: 636–653.

Calvet, N., Hartmann, L. W., Kenyon, S. J., and Whitney, B. A. 1994. Flat spectrum T Tauri stars: The case for infall. *Astrophysical Journal* 434: 330–340.

Cassen, P., and Moosman, A. 1981. On the formation of protostellar disks. *Icarus* 48: 353–376.

Chabrier, G. 2003. Galactic Stellar and Substellar Initial Mass Function. *Publications of the Astronomical Society of the Pacific* 115: 763–795.

Chandrasekhar, S. 1939. *Stellar Structure* (New York: Dover).

Chiang, E. I., and Goldreich, P. 1997. Spectral energy distributions of T Tauri stars with passive circumstellar Disks. *Astrophysical Journal* 490: 368–376.

D'Alessio, P., Cantö, J., Calvet, N., and Lizano, S. 1998. Accretion disks around young objects. I. The detailed vertical structure. *Astrophysical Journal* 500: 411–427.

Duchêne, G., and Kraus, A. 2013. Stellar multiplicity. *Annual Review of Astronomy and Astrophysics* 51: 269–310.

Goldreich, P., and Sridhar, S. 1995. Toward a theory of interstellar turbulence. 2: Strong alfvenic turbulence. *Astrophysical Journal* 438: 763–775.

Goodman, A. A. 1990. Interstellar magnetic fields: An observational perspective. Ph.D. Thesis, Harvard University.

Haisch, K. E., Lada, E. A., and Lada, C. J. 2001. Disk frequencies and lifetimes in young clusters. *Astrophysical Journal* 553: L153–156.

Hayashi, C. 1981. Structure of the solar nebula, growth and decay of magnetic fields and effects of magnetic and turbulent viscosities on the nebula. *Progress of Theoretical Physics Supplement* 70: 35–53.

Heiles, C. H., Goodman, A. A., McKee, C. F., and Zweibel, E. G. 1993. Magnetic fields in starforming regions: Observations. In Levy, E., and Mathews, M. S., (Eds.), *Protostars and Planets III* (Tucson: University of Arizona Press), pp. 279–326.

Houlahan, P., and Scalo, J. 1992. Recognition and characterization of hierarchical interstellar structure. II. Structure tree statistics. *Astrophysical Journal* 393: 172–187.

Hoyle, F. 1960. On the origin of the solar system. *Quarterly Journal of the Royal Astronomical Society* 1: 28–55.

Jijina, J.; Myers, P. C.; Adams, Fred C. 1999. Dense Cores Mapped in Ammonia: A Database. Astrophysical Journal Supplement 125: 161–236

Johansen, A., Oishi, J. S., Mac Low, M.-M., Klahr, H., Henning, T., and Youdin, A. 2007. Rapid planetesimal formation in turbulent circumstellar disks. *Nature* 448: 1022–1025.

Kant, I. 1755. *Allegmeine Natuigeschichte and Theorie des Himmels*. Germany. English Translation: Kant, I. 1986. *Universal Natural History and Theory of the Heavens*. Translate by Jaki, S. K. (Edinburgh: Scottish Academic Press).

Kenyon, S. J., Calvet, N., and Hartmann, L. W. 1993. The embedded young stars in the Taurus-Auriga molecular cloud. I. Spectral energy distributions. *Astrophysical Journal* 414: 676–694.

Kenyon, S., and Hartmann, L. 1987. Spectral energy distributions of T Tauri stars: Disk flaring and limits on accretion. *Astrophysical Journal* 323: 714–733.

Königl, A., and Pudritz, R. 2000. Disk winds and the accretion-outflow connection. In Mannings, V. Boss, A. P., and Russell S. S. (Eds.), *Protostars and Planets IV* (Tucson: University of Arizona Press), pp. 759.

Kroupa, P. 2001. On the variation of the initial mass function. *Monthly Notices of the Royal Astronomical Society* 322: 231–246.

Lada, C. J. 1985. Cold outflows, energetic winds, and enigmatic jets around young stellar objects. *Annual Review of Astronomy and Astrophysics* 23: 267–317.

Lada, C. J., and Shu, F. H. 1990. The formation of sunlike stars. *Science* 1111: 1222–1233.

Ladd, E. F., Adams, F. C., Casey, S., Davidson, J.A., Fuller, G. A., Harper, D. A., Myers, P. C., and Padman, R. 1991. Far infrared and submillimeter

wavelength observations of star forming dense cores. II. Spatial distribution of continuum emission. *Astrophysical Journal* 382: 555–569.
Laplace, P. S. 1796. *Exposition du Systeme du Monde* (Paris). English Translation in: Knickerbocker, W. S. 1927. *Classics of Modern Science* (Boston: Beacon Press).
Larson, R. B. 1969. Numerical calculations of the dynamics of collapsing protostar. *Monthly Notices of the Royal Astronomical Society* 145: 271–295.
Larson, R. B. 1981. Turbulence and star-formation in molecular clouds. *Monthly Notices of the Royal Astronomical Society* 194: 809–826.
Laughlin, G. P., and Bodenheimer, P. 1994. Nonaxisymmetric evolution in protostellar disks. *Astrophysical Journal* 436: 335–354.
Laughlin, G. P., Bodenheimer, P., and Adams, F. C. 2004. The core accretion model predicts few jovian-mass planets orbiting red dwarfs. *Astrophysical Journal* 612: L73–76.
Liist, R. 1952. Die entwicklung einer um einen zeutralkorper rotierenden gasmasse. I. Loesungen de hydrodynamischen gleichungenmit turulenter reibung. *Zeitschrift fur Naturforschung* 7a: 87–98.
Lin, D. N. C., and Papaloizou, J. C. B. 1985. On the dynamical origin of the solar system. In Black, D. C., and Mathews, M. S. (Eds.), *Protostars and Planets II* (Tucson: University of Arizona Press), pp. 981–1072.
Lin, D. N. C., and Papaloizou, J. C. B. 1986a. On the tidal interaction between protoplanets and the primordial solar nebula. II. Self-consistent nonlinear interaction. *Astrophysical Journal* 307: 395–409.
Lin, D. N. C., and Papaloizou, J. C. B. 1986b. On the tidal interaction between protoplanets and the primordial solar nebula. III. Orbital migration of protoplanets. *Astrophysical Journal* 309: 846–857.
Lizano, S., and Shu, F. H. 1989. Molecular cloud cores and bimodal star formation. *Astrophysical Journal* 342: 834–854.
Lynden-Bell, D., and Pringle, J. E. 1974. The evolution of viscous disks and the origin of the nebular variables. *Monthly Notices of the Royal Astronomical Society* 168: 603–637.
Mac Low, M.-M., and Klessen, R. S. 2004. Control of star formation by supersonic turbulence. *Reviews of Modern Physics* 76: 125–194.
Marois, C., Macintosh, B., Barman, T., Zuckerman, B., Song, I., Patience, J., Lafreniére, D., and Doyon, R. 2008. Direct imaging of multiple planets orbiting the star HR 8799. *Science* 322: 1348–1352.
Mayor, M., and Queloz, D. 1995. A Jupiter-mass companion to a solar-type star. *Nature* 378: 355–359.
McKee, C. F., and Ostriker, E. C. 2007. Theory of star formation. *Annual Review of Astronomy and Astrophysics* 45: 565–687.
Miller, G. E., and Scalo, J. M. 1979. The initial mass function and stellar birth rate in the solar neighborhood. *Astrophysical Journal Supplement* 41: 513–547.

Mouschovias, T. 1976. Nonhomologous contraction and equilibria of self-gravitating magnetic interstellar clouds embedded in an intercloud medium: Star formation. I. Formulation of the problem and method of solution. *Astrophysical Journal* 206: 753–767.

Mouschovias, T., and Spitzer, L. 1976. Note on the collapse of magnetic interstellar clouds. *Astrophysical Journal* 210: 326–327.

Myers, P. C. 1985. Molecular cloud cores. In Black, D. C., and Mathews, M. S., (Eds.), *Protostars and Planets II* (Tucson: University of Arizona Press), pp. 81–103.

Myers, P. C., Fuller, G. A., Mathieu, R. D., Beichman, C. A., Benson, P. J., Schild, R. E., and Emerson, J. P. 1987. Near-infrared and optical observations of IRAS sources in and near dense cores. *Astrophysical Journal* 319: 340–357.

Myers, P. C., and Fuller, G. A. 1992. Density structure and star formation in dense cores with thermal and nonthermal motions. *Astrophysical Journal* 396: 631–648.

Myers, P. C., and Goodman, A. A. 1988. Magnetic molecular clouds: Indirect evidence for magnetic support and ambipolar diffusion. *Astrophysical Journal* 329: 392–405.

Nakano, T. 1984. Contraction of magnetic interstellar clouds. *Fundamentals of Cosmic Physics* 9: 139–232.

Palla, F., and Stahler, S. W. 1990. The birthline for intermediate mass stars. *Astrophysical Journal* 360: L47–L50.

Phillips, A. C. 1994. *The Physics of Stars* (New York: Wiley).

Pringle, J. E. 1981. Accretion discs in astrophysics. *Annual Review of Astronomy and Astrophysics* 19: 137–162.

Quintana, E. V., Barclay, T., Raymond, S. N., *et al.* 2014. An Earth-sized planet in the habitable zone of a cool star. *Science* 344: 277–280.

Rafikov, R. R. 2005. Can giant planets form by direct gravitational instability? *Astrophysical Journal* 621: L69–72.

Rathborne, J. M., Lada, C. J., Muench, A. A., Alves, J. F., Kainulainen, J., and Lombardi, M. 2009. Dense cores in the pipe nebula: An improved core mass function. *Astrophysical Journal* 699: 742–753.

Rice, T. S., Goodman, A. A., Bergin, E. A., Beaumont, C., and Dame, T. M. 2016. A Uniform catalog of molecular clouds in the Milky way. *Astrophysical Journal* 822: 52

Rucinski, S. M. 1985. IRAS observations of T Tauri and post-T Tauri stars. *Astronomical Journal* 90: 2321–2330.

Ruden, S. P., and Lin, D. N. C. 1986. The global evolution of the primordial solar nebula. *Astrophysical Journal* 308: 883–901.

Rydgren, A. E., and Zak, D. S. 1987. On the spectral form of the infrared excess component in T Tauri systems. *Publications of the Astronomical Society of the Pacific* 99: 141–145.

Salpeter, E. E. 1955. The luminosity function and stellar evolution. *Astrophysical Journal* 121: 161–167.

Scalo, J. M. 1986. The stellar initial mass function. *Fundamentals of Cosmic Physics* 11: 1–278.

Scalo, J. M. 1987. Theoretical approaches to interstellar turbulence. In Hollenbach, D. J., and Thronson, H. A. (Eds.), *Interstellar Processes* (Dordrecht: Reidel), pp. 349–392.

Shang, H. 2007. Jets and molecular outflows from Herbig-Haro objects. *Astrophysics and Space Science* 311: pp. 25–34.

Shu, F. H. 1977. Self-similar collapse of isothermal spheres and star formation. *Astrophysical Journal* 214: 488–497.

Shu, F. H. 1983. Ambipolar diffusion in self-gravitating isothermal layers. *Astrophysical Journal* 273: 202–213.

Shu, F. H. 1985. Star formation in molecular clouds. In van H. Woerden, Burton, W. B., and Allen, R. J., (Eds.), *The Milky Way* (IAU Symposium No. 106) (Dordrecht: Reidel), pp. 561–565.

Shu, F. H., Adams, F. C., and Lizano, S. 1987. Star formation in molecular clouds: Observation and theory. *Annual Review of Astronomy and Astrophysics* 25: 23–81.

Shu, F. H., Lizano, S., and Adams, F. C. 1987. Star formation in molecular cloud cores. In Peimbert, M., and Jugaku, J. (Eds.), *Star Forming Regions (IAU Symposium No. 115)* (Dordrecht: Reidel), pp. 417–434.

Shu, F. H., Lizano, S., Ruden, S. P., and Najita, J. 1988. Mass loss from rapidly rotating magnetic protostars. *Astrophysical Journal* 328: L19–L23.

Shu, F. H., Najita, J., Wilkin, F., Ruden, S. P., and Lizano, S. 1994. Magnetocentrifugally driven flows from young stars and disks. I. A generalized model. *Astrophysical Journal* 429: 781–796.

Shu, F. H., Tremaine, S., Adams, F. C., and Ruden, S. P. 1990. SLING amplification and eccentric gravitational instabilities in gaseous disks. *Astrophysical Journal* 358: 495–514.

Silk, J. 1995. A theory for the initial mass function. *Astrophysical Journal* 438: L41–L44.

Stahler, S. W. 1983. The birthline of low-mass stars. *Astrophysical Journal* 274: 822–829.

Stahler, S. W. 1988. Deuterium and the stellar birthline. *Astrophysical Journal* 332: 804–825.

Stahler, S. W., Shu, F. H., and Taam, R. E. 1980. The evolution of protostars. I. Global formulation and results. *Astrophysical Journal* 241: 637–654.

Stepinsky, T. F., and Levy, E. H. 1990. Dynamo magnetic field induced angular momentum transport in protostellar nebulae: The "minimum mass protosolar nebula." *Astrophysical Journal* 350: 819–826.

Terebey, S., Shu, F. H., and Cassen, P. 1984. The collapse of the cores of slowly rotating isothermal clouds. *Astrophysical Journal* 286: 529–551.

Walker, C. K., Adams, F. C., and Lada, C. J. 1990. 1.3 millimeter continuum observations of cold molecular cloud cores. *Astrophysical Journal* 349: 515–528.

Winn, J. N., and Fabrycky, D. C. 2015, The occurrence and architecture of exoplanetary systems. *Annual Review of Astronomy and Astrophysics*, 53: 409–447.

Wood, D. O. S., Myers, P. C., and Daugherty, D. A. 1994. IRAS images of nearby dark clouds. *Astrophysical Journal Supplement* 95: 457–501.

Woodward, J. W., Tohline, J. E., and Hashisu, I. 1994. The stability of thick self-gravitating disks in protostellar systems. *Astrophysical Journal* 420: 247–267.

Young, C. H., and Evans, N. J. 2005. Evolutionary signatures in the formation of low-mass protostars. *Astrophysical Journal* 627: 293–309.

Zinnecker, H., McCaughrean, M. J., and Wilking, B. A. 1993. The initial stellar population. In Levy, E., and Mathews, M. S. (Eds.), *Protostars and Planets III* (Tucson: University of Arizona Press), pp. 429–496.

Zinnecker, H., and Yorke, H. W. 2007. Toward understanding massive star formation. *Annual Review of Astronomy and Astrophysics* 45: 481–563.

Zuckerman, B., and Evans, N. J. II. 1974. Models of massive molecular clouds. *Astrophysical Journal Letters* 192: L149–152.

Zuckerman, B., and Palmer, P. 1974. Radio emission from interstellar molecules. *Annual Review of Astronomy and Astrophysics* 12: 279–313.

Zuckerman, B., Forveille, T., and Kastner, J. H. 1995. Inhibition of giant-planet formation by rapid gas depletion around young stars. *Nature* 373: 494–496.

Chapter 6

The Origin and Evolution of Life in the Universe

Christopher P. McKay

Introduction

Are we alone? We wonder whether life on Earth is a rare, even unique, phenomenon or whether life is to be found in myriad forms on many worlds throughout the Universe. As human beings, we ponder the probability that life elsewhere has developed intelligence to levels comparable with ours; and if so, how can we communicate with these life forms? The question of life in the Universe necessarily straddles scientific fields from astronomy to zoology — fields that appear at first to have no common thread.

The discovery of planets orbiting other stars shows that there are numerous planets in our own galaxy. We expect that life could be found on virtually all that are Earth-like in their size and distance from a central star. The first test of our theories for the distribution of life beyond the Earth will be a search for life on the other worlds of our solar system and the search for indirect evidence of life on **exoplanets**. The question of advanced life and, in particular, technological life elsewhere is more difficult to address and the only strategy for presently seeking such life is to listen for messages from alien transmitters.

Life begins at home, and so this chapter surveys the origin and evolution of life by reviewing what we have learned from life on Earth. From this single point of data, one can extrapolate outward to the stars and attempt to place the history of life on Earth in the context of the history of the Universe.

Life on Earth

Despite its various forms, there is only one example of life on Earth. One of the most profound results of modern science is the unity of biochemistry. All life forms on Earth have the same basic elemental makeup, use a common set of biochemical building blocks, and share a common ancestor and genetic code.

Life on Earth is made of common elements. If we use the well-studied gut bacteria, *Escherichia coli* (*E. coli*) as our model, then the important elements for life, making up over 99% of the total by number, are H, O, C, and N. These elements are among the most abundant in the solar system and in the galaxy. The next tier of elements in life P, Na, K, S, Ca, and Mg, together sum to less than 1%. Table 6.1 give the list of elements in *E. coli* as determined from the values in Davies and Koch (1991).

The dominance of H and O, and their relative amounts, reflects the importance of H_2O in living systems.

Life uses the elements H, O, C, and N to make elaborate biochemicals. This fact is perhaps best illustrated by the **amino acids** from which all **proteins** are made. Of all possible amino acids, proteins are built primarily from a common set of 20. These are shown in Figure 6.1, along with the other basic biochemicals of life (Lehninger, 1975). Amino acids come in left- and right-hand versions. The handedness refers to the relative

Table 6.1. Elements in *E. coli* based on Davies and Koch (1991).

Element	Percent by number
H	63
O	26
C	7.8
N	1.9
P	0.17
Na	0.11
K	0.07
S	0.06
Ca	0.04
Mg	0.02

The Origin and Evolution of Life in the Universe

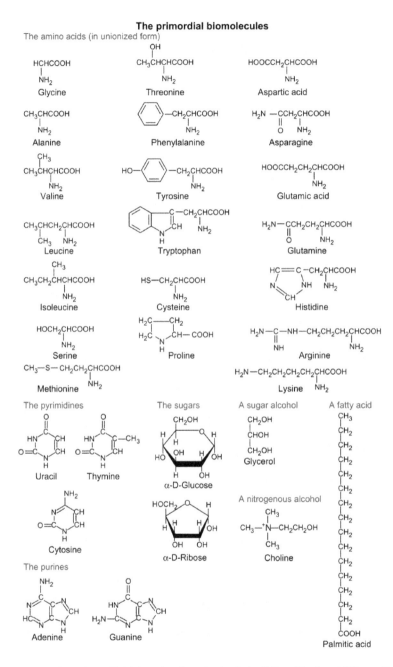

Figure 6.1. The primordial biomolecules: the basic building blocks of life on Earth. (Reprinted by permission from Lehninger, 1975.)

location of subcomponents of a given amino acid. In nonbiological processes, amino acids are found roughly equally in the left- and right-hand versions; in biology, only left-hand amino acids are used for making proteins.

Another universal characteristic of all life is that it shares the same genetic code instantiated in **RNA** and **DNA**. The five nucleic acid subunits of DNA and RNA, the 20 amino acids, and a few other biomolecules are the basic building blocks of biology. Life strings these together to form the long chains needed for structure, genetics, catalysis, and all the other functions in a cell. All life on Earth, from the *E. coli* bacteria to the blue whale, is constructed from this same set of building blocks.

The similarity between the genetic material of all life implies another important property of life on Earth; it shares a common ancestor. Thus, it is possible to construct a diagram — known as a **phylogenetic tree** — that shows how all lifeforms are related. Such a phylogenetic tree is based on variations in some molecule that is common to all life forms, typically a type of RNA (Woese, 1987). If the rate of variation in the RNA being tracked can be correlated with time, then branching points on the phylogenetic tree can be assigned dates. These dates can be correlated with environmental events indicated in the geological record. Following this process, the best estimate for the date of the common ancestor is about 3.5 billion years ago (e.g., Cavalier-Smith, 2010). The common ancestor was not the origin of life and not necessarily even the first cell. On the contrary, the common ancestor almost certainly represented a considerable period of evolution after the origin of life. There are three possible explanations for any specific features of the common ancestor. One possibility is that the features of the common ancestor may have been crucial to the origin of life or to its early evolution — it does represent how life started. Alternatively, early life on Earth may have been stressed by some ecological disaster, such as a large comet impact, that exterminated all life except some microorganisms living in a protected environment with the features of the common ancestor. After the event, these organisms recolonized the planet. A third possibility is that the nature of the common ancestor may also be purely a result of chance and be unconnected in any fundamental way to the events on the early Earth.

A schematic of the phylogenetic tree of life on Earth is shown in Figure 6.2. The common ancestor is indicated at the base of the tree and the three Domains of life are shown: the archaea, the bacteria, and the eukarya. Note that viruses — which are clearly alive and share the common biochemistry with the rest of life — are not in the tree shown in Figure 6.2. This is

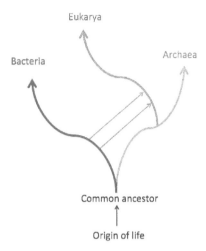

Figure 6.2. Schematic diagram of the phylogenetic tree of life on Earth based on similarities in conserved RNA. The bottom of the tree is the common ancestor. The main branches are the bacteria, the archaea, and the eukarya. The eukarya is formed by a symbiosis of an archaea host and repeated bacteria endosymbionts.

because viruses are degenerated parasites that do not have the full biochemical set that free-living cells contain. Thus, any particular DNA, RNA, or protein used for phylogenetic comparisons will not be contained by all viruses as some contain DNA but not RNA and vice versa, for example.

From the base of the common ancestor, the phylogenetic tree splits into the two domains of life that contain only microscopic life forms: bacteria and archaea. These microorganisms differ from each other in subtle biochemical ways and were simply classified as a a single group, bacteria, until Woose (1987) discovered that they were phylogenically distinct. The eukarya comprise what we refer to as life in the common use of the term. Multicellular life is found only in the eukarya and virtually all use oxygen for metabolism. This has led to the suggestion that multicellular life anywhere in the Universe will require the availability of O_2 (Catling et al. 2005).

The origin of life on Earth remains a puzzle. A timeline (see Figure 6.3) shows that life appears early in Earth's history. The formation of the Earth was completed about 3.8 billion years ago when the last of the **planetesimals** that formed the planets were accreted. This event, known as the **heavy bombardment**, is recorded in the densely cratered surfaces of the Moon, Mercury, and Mars. On Earth, the craters caused by the heavy bombardment have long been eroded.

Origin and Evolution of the Universe

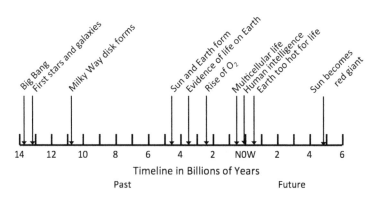

Figure 6.3. Timeline showing the main events in the history of the Universe and the history of life on Earth and its future through the next 5 billion years (assuming humans do not alter the Earth or the solar system).

There is definite evidence for life on Earth 3.5 billion years ago in the form of **fossil** microbial mats (Tice and Lowe, 2004; Allwood et al. 2006). Furthermore, the nature of this life indicates that **photosynthesis** was present and formed the basis of complete microbial communities similar to microbial mats found in many locations on Earth today. Thus, there was certainly comparatively sophisticated microbial life by 3.5 billion years ago. Moreover, there is indirect evidence that life on Earth may have existed even earlier. Life, photosynthesis in particular, produces a shift in the relative amounts of the **isotopes** of carbon. Compared with the atmosphere, carbon within living systems is about 2% enriched with the lighter isotope, ^{12}C, over with the heavier isotope, ^{13}C. This shift is seen in the sedimentary material of biological origin for the past 3.5 billion years. Interestingly, this shift is also seen in **sedimentary deposits** that are 3.8 billion years old (Schidlowski, 1988; Abramov and Mojzsis, 2009) — during the end of the heavy bombardment. This information is consistent with photosynthetic life at that time, but it is not proof of life because the sediments are too altered to preserve any direct fossils. Nonetheless, the evidence from Earth indicates that life arose very rapidly and possibly instantaneously (which, geologically speaking, is a period of approximately several tens of millions of years). A possible extrapolation of this observation is that on any planet with conditions similar to those on early Earth, life would quickly emerge, even as the planet was forming.

It is interesting to consider what evidence we have that the life on Earth was the same 3.5 billion years ago as it is today. Has there been a continuity of "life-as-we-know-it" over Earth history? We cannot determine the biochemistry of ancient fossils and hence the only indication of the continuity of life on Earth is the approximate constancy of the ^{13}C isotope shift in organic carbon recorded in the sedimentary record.

Although we know that life appeared early in the history of Earth, we do not know the details of how life originated. The most widely held theory for the origin of life is based on the original suggestions of A. I. Oparin and J. B. S. Haldane. Independently, they suggested that life originated from **organic material** that had been produced abiologically. This theory for the origin of life received convincing support in the experiments of Stanley Miller. In these experiments, and in subsequent works along these lines, gas mixtures thought to represent the plausible composition of the atmosphere of early Earth were subjected to electrical discharges (Miller, 1953).

These experiments are remarkable in two ways. First, the fact that organic molecules are so readily made from inorganic, albeit **reducing**, gases suggests that organic chemistry, far from being the sole purview of living systems, should be commonplace in the cosmos. Second, the organics produced in the **abiotic synthesis** were not a mere random collection of molecules but were composed of many of the same compounds that are found in life. The chemistry of life appears, therefore, to be a part of the cosmic organic chemistry.

This overall view of cosmic organic chemistry received considerable support by the subsequent discovery of organics in space. Organics were first discovered in meteorites and the organics present were similar to those produced in Miller's abiotic syntheses. Organics were then discovered (e.g., Ehrenfreund and Charnley, 2000) in the interstellar medium, in comets, in the outer solar system, on Titan, and, especially in the plume of Enceladus (Waite *et al.*, 2009). In fact, only the low levels of organics on Mars (Freissinet *et al.*, 2015) is incongruous with a prevailing organic chemistry in the solar system and the cosmos. Organics on the surface of Mars appear to have been destroyed by the oxidants produced by **ultraviolet** (UV) light and cosmic radiation.

Although the evidence for the widespread abiotic origin of organics relevant to biochemistry is compelling, the standard theory for the origin of life suffers from a major flaw — it has not been possible to create life from abiotic organics. This is a more serious issue than one might imagine at first. Lazcano and Miller (1994) have suggested that the time required for the emergence of life from the prebiotic soup may have been short (on geological timescales) — less than 10 million years. If this is true, a factor of 100 million enhancement in the experimental process compared with the process in nature should produce life in the laboratory in 1 year. However, it is interesting to note that Orgel (1998) criticized this conclusion and stated that we do not understand the steps that lead to life and therefore we cannot estimate the time required: "Attempts to circumvent this essential difficulty are based on misunderstandings of the nature of the problem." Thus, until new data are obtained the problem of origin of life remains unsolvable. Indeed, even the common assumption that life originated on Earth is not directly supported, or refuted, by data.

Even setting aside the optimistic view that life only required 10 million years to begin, when one compares random processes in nature with directly controlled experiments, it is not unreasonable to expect that rate enhancements of 100 million are possible. Thus, it is problematical that, 60 years after the first experiment, life remains to be created in abiotic simulations, although some incremental knowledge of organic chemistry has been obtained (McCollom, 2013).

There are alternate theories for the origin of life (Davis and McKay, 1996). Some postulate that life came to Earth from elsewhere. This notion, called **panspermia**, has been gaining attention as researchers appreciate the short interval for the origin of life on Earth and as mounting evidence shows how impacts can launch materials from one planet to another in a way that could preserve dormant spores (Melosh, 1988). Other theories for the origin of life allow a terrestrial origin but differ from the standard theory in that they do not postulate an organic origin for life. Instead, they postulate that the first life forms were composed of clay minerals that evolved into organic-based life later (Cairns-Smith, 1982). Although the standard theory for the origin of life assumes that the first lifeforms lived by consuming organic material already present in the environment, a process known as heterotrophy, other theories suggest that the first life form could have derived its energy from sunlight or even from chemical reactions (such as $4H_2 + CO_2 = CH_4 + 2H_2O$). Even if life did arise from a **primordial soup** and stayed alive by consuming that soup, the question of where the ingredients for the soup came from still remains.

The original work of Miller and others suggested that the organics in the soup were produced on Earth in an atmosphere rich in methane and ammonia — considerably different from our present atmosphere. However, methane and ammonia may not have been stable in Earth's early atmosphere because of destruction by UV light from the Sun, which has led to the suggestion that the ingredients for the prebiotic soup were carried to a receptive Earth by comets and meteorites (Ehrenfreund and Charnley, 2000; Pizzarello, 2004). One common feature of all theories for the origin of life on Earth is that life required liquid water environments (Davis and McKay, 1996). The diversity of theories for the origin of life illustrates how uncertain we are about the events that led to life on Earth. What we can say for certain is that life appeared, if not originated, early in Earth's history and required liquid water.

An interesting biochemical development that bears on the origin and early development of life is the discovery that RNA can act both to store genetic information (such as DNA) and to affect biochemical reactions (such as enzymes, which are proteins) (Gilbert, 1986). In modern life forms, genetic information is recorded in DNA but, before the information can be used, it must be transcribed into proteins, the working molecules of biochemistry. The separation of biochemistry into information molecules and action molecules was long a puzzle. Because the molecules were mutually interdependent, which originated first was unclear. The discovery that RNA is simultaneously capable of both functions appears to provide one possible solution to this dilemma. Thus, an early stage in the evolution of life may have involved a so-called RNA world (Robertson and Joyce, 2012). Later, evolution led to the development of more efficient genetic material (DNA) and more efficient enzymes (proteins) and formed the basis for the DNA world we have now. The timing of this evolution is uncertain but it would appear to have occurred before the oldest fossils, at 3.5 billion years ago. The DNA world must have occurred before the divergence of life into the three main groups shown in Figure 6.2, because comparison of the genetic difference between these groups of life implies that the genetic code is 3.8 ± 0.6 billion years old (Eigen *et al.*, 1989).

Although its origin is shrouded in mystery, life has left a more complete account of its subsequent history in the geological record. We know that life on Earth remained microbial for about 3 billion years (Figure 6.3). For about the first billion years of this time period, the atmosphere of Earth did not contain any appreciable oxygen. Although, almost certainly, photosynthetic microorganisms were producing oxygen, it was consumed by natural sources of reducing material, including sulfide from volcanoes, dissolved iron in the oceans, and organic material. A net production of oxygen is achieved only when the organic matter produced in photosynthesis is secured in sediments. Exposed organic material oxidizes and thus consumes the oxygen that was created when the organic matter was produced. Not until about 2 billion years ago did oxygen production overcome these losses so that oxygen could begin to accumulate in the atmosphere. The rate of buildup of atmospheric oxygen is uncertain, but probably continued until 600 million years ago (Figure 6.3).

Multicellular life developed rapidly after the rise of oxygen. Oxygen may have been critical to the development of multicellular life in two ways. First, an ozone layer, formed after oxygen accumulated, may have been needed to protect multicellular life from the harmful UV light. Second, and more probable, oxygen may have been needed to provide a metabolic reaction powerful enough to supply the needs of large multicellular organisms; all multicellular life forms breathe oxygen.

The buildup of oxygen on Earth is one of the key events in the history of life on this planet. Oxygen production was dominated by biological photosynthesis, but geological processes controlled the rate at which organic material was sequestered and the rate at which reduced gases were vented to the atmosphere. Thus, the timing of an oxygen-rich atmosphere and the concomitant development of multicellular life was set by the geological nature of Earth (Knoll, 1985; Lyons *et al.*, 2014).

Microbial life on Earth has adapted to many unusual conditions, and the distribution of microorganisms tells us the factors that are truly limiting for life. Microorganisms can thrive in boiling water, in acid, and in concentrated salt solutions. Other microorganisms live on the bottom of the sea floor in **hydrothermal vents** where they consume H_2S by reacting with O_2 dissolved in seawater (the ultimate source of this O_2 is photosynthesis on the surface of Earth). Many microbes live completely without O_2 and, in fact, oxygen is a deadly poison to them. These anaerobic bacteria are responsible for the degradation of waste material in landfills and sewage. However, the one environmental requirement that all life on Earth shares is the requirement for liquid water. Some organisms can survive dormant in a dried state, but to grow all life on Earth needs liquid water (Kushner, 1981). Ice will not do. Life grows in ice and snow only when there is a liquid fraction. Vapor will not do, although some organisms, notably **lichens**, can concentrate vapor to liquid as long as the **relative humidity** is above 70%. In addition to liquid water, the requirements for life are energy, carbon, and some other elements listed in Table 6.2 along with an indication of how commonly these requirements are satisfied on the other planets of our solar system. It is interesting to note that carbon,

Table 6.2. Requirements for life (adapted from McKay, 1991).

Requirement	Occurrence in the solar system
Energy as sunlight or as chemical oxidation-reduction reactions	Common
Carbon	Common, as CO_2 and CH_4
Liquid water	Rare
N, P, S, and other elements	Common

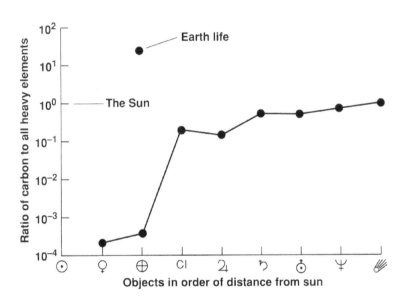

Figure 6.4. Carbon abundance is the solar system: Ratio of carbon atoms to total heavy atoms (heavier than He) for various solar system objects, which illustrates the depletion of carbon in the inner solar system. The abscissa is not a true distance scale, but the objects are ordered by increasing distance from the Sun. Mars is not shown because the size of its carbon reservoir is unknown. The symbols, from Left to Right, are: Sun, Venus, Earth, the most common class of asteroids, Jupiter, Saturn, Uranus, Neptune, and comets. (From McKay, 1991).

a key element for life is actually depleted on Earth compared to the rest of the solar system (Figure 6.4). But because it has common forms that are gaseous (CO_2 and CH_4), carbon is adequately present at the surface of the Earth.

The Search for Life Beyond the Earth

In computer programming, there is a rule known as the zero-one-infinity rule. This rule states that the logical number of any item is either zero, one, or infinity — where infinity means some large number set by some external parameter such as the total size of the computer, etc. Issac Asimov applied this logic to items in the Universe in his book *The Gods Themselves*. We know that for galaxies, stars, and planets the number is infinity. For life, we know that the number is at least one. Thus, the key goal in astrobiology is the search for a second genesis of life (McKay, 2001). Showing that life has independently appeared in the Universe twice moves biology from the category of one to that of infinity. Of course intelligent life capable of communication would be preferred, but even a microbial second genesis would suffice. Where can we find a second genesis of life? How do we search? The start of the search is in our own Solar System (McKay, 2004) and is based on what we know of life on Earth.

Liquid water is the defining and quintessential requirement for life on Earth. Therefore, the search for liquid water is an operational approach to the search for life elsewhere. In our own Solar System, Mars is of prime interest because of evidence that liquid water once flowed on the Martian surface. The orbital images showing fluvial features are a direct indication that a liquid — water is the only plausible candidate — flowed on the Martian surface in vast quantities (see Figure 6.5).

In 1976, two Viking spacecrafts landed on Mars and began a search for microscopic life forms in the soil. The Viking biology package was composed of three instruments, each of which conducted experiments on a sample of the Martian surface. One experiment searched for photosynthesis, and the other two looked for organisms capable of consuming a nutrient solution — a soup-like broth. The results were intriguing. In both experiments with the nutrient solution, gases were released, which indicated some sort of activity. Each of the two Viking landers also carried an instrument for the characterization of organic material, a **gas chromatograph mass spectrometer** (GCMS) — no Martian organics were reported.

These Viking results were puzzling, both the gases released from the soil and the lack of organics. Organics should be present on Mars just from meteorite infall even in the absence of any biological production. The resolution of these puzzles emerged only in 2008 with the discovery of high levels of perchlorate in the soil of Mars by the Phoenix mission

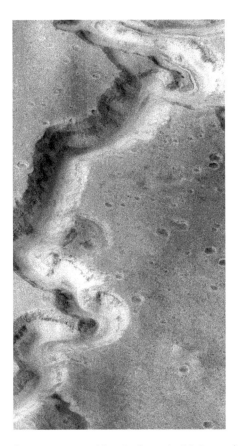

Figure 6.5. Evidence for past water on Mars is shown in this image from orbit of Nanedi Vallis. The scene covers an area 9.8 km by 18.5 km; the canyon is about 2.5 km wide. This image is the best evidence we have of surface liquid water anywhere outside Earth. *Source*: Photo from NASA/Malin Space Sciences.

(Hecht *et al.*, 2009). Navarro-González *et al.*, (2010) showed that when heated in the *GCMS* the perchlorate would become reactive and destroy any organics present in the sample producing chlorinated organics as was seen by the Viking landers — effectively explaining the lack of detection of Martian organics by the Viking instruments. Quinn *et al.*, (2013) showed that **cosmic rays** reaching the surface of Mars would create reactive products from the perchlorate which could explain the reactivity seen in the Viking nutrient experiments. The combined effects of cosmic radiation and perchlorate indicate that evidence for past life on Mars will have to be sought deep (5–10 meters) below the surface. The Viking results notwithstanding, past life on Mars remains a compelling possibility — and a strong motivation for human exploration.

Geomorphological evidence suggests that liquid water existed on the surface of Mars at approximately the time that the first life appeared on Earth, between 3.8 and 3.5 billion years ago (McKay, 1986; McKay and Stoker, 1989). The possibility of the origin of life on Mars is based on analogy with Earth. All the major habitats and microenvironments that would have existed on Earth during the formation of life would have been expected on early Mars as well: hot springs, salt pools, rivers, lakes, volcanoes, and so forth. Even tidal pools would have existed on Mars, albeit at a much reduced level because there would have been only solar tides. The possible nonbiological sources of organic material would have supplied both planets. Perhaps the major unknown is the duration of time that Mars had Earth-like environments compared with the time required for the origin of life. The length of neither of these times is known precisely, but current theories suggest that the lengths may be comparable (McKay and Davis, 1991).

The events that led to the origin of life on early Earth and that may have also occurred on early Mars may be better preserved in the sediments on Mars than those on Earth. On Earth, sediments that date back 3.5 to 4.0 billion years ago are rare, and those that exist are usually severely altered. On Mars, over half the planet dates back to the end of heavy bombardment, about 3.8 billion years ago, and has been well-preserved at low temperatures and pressures. Grotzinger *et al.* (2015) report on rover studies of the sediments and implied paleoclimate of an ancient lake deposit in Gale Crater on Mars. Thus, although there may be no life on Mars today, Mars may hold the best record of the events that led to the origin of life on Earth-like planets.

Beyond Mars, there are other worlds which might hold evidence of a second genesis of life. In particular, Europa, a moon of Jupiter, and Enceladus, a moon of Saturn, both contain ice-covered oceans. Enceladus is of particular interest because there is a plume originating from the ocean venting into space (shown in Figure 6.6). The plume contains organic material, nitrogen in forms that are biologically useful, and chemical energy sources that could be used by life. In addition, there is evidence of salts in the plume. Thus, the plume on Enceladus seems to originate from an ocean that is habitable (McKay *et al.* 2014). Samples of life may be streaming into space making the plume an attractive target for collection and analysis.

The Origin and Evolution of Life in the Universe

Figure 6.6. Side-on view of the spectacular plumes venting from under the ice surface of Saturn's moon Enceladus. This photograph from a NASA spacecraft shows the icy particles and vapor backlighted by sunlight. The small dark object to the left center of the image indicates how far above Enceladus's surface the Cassini spacecraft passed when if flew through the plumes on October 28, 2015.

Titan, the largest moon of the planet Saturn, possesses a unique opportunity and challenge to our understanding of life. Titan is the only world we know other than Earth that has a liquid on its surface, but that liquid is a methane mixture at a temperature of 95 K. Titan's thick atmosphere is composed mostly of nitrogen with a few percent methane and a thick organic haze. There are lakes, rain, and clouds of methane and ethane. The suggestion that there may be life in the liquid methane on Titan challenges our concept of liquid water-based life. But the discovery of any such biology would be our first compelling indication that life in the Universe is considerably more diverse than the one example we find on Earth.

Exoplanets

In the past 20 years, there has been a vast expansion in the number of known planets detected around other stars. The largest share of these were discovered by NASA's Kepler spacecraft by finding **transits** (eg. Borucki *et al.* 2013). Some of these planets are similar to Earth in size and inferred mass, and orbit their star at a distance that would be considered habitable. An important discovery is that rocky extrasolar planets have bulk elemental compositions similar to those of Earth and that many extrasolar minor planets have crusts, mantle, and cores — this finding has important implications for the origin of our own Solar System (Zuckerman and Young, 2018).

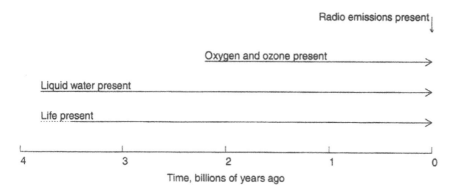

Figure 6.7. Timeline showing methods for detecting life and habitable conditions on Earth from a nearby star. Liquid water has always been an indicator of habitability on Earth. Oxygen has been present for about half of Earth's biological history; radio emissions have been present for only the past century. A similar approach could be used from Earth to detect life and liquid water on exoplanets and *exomoons*.

If life is present on any of the exoplanets discovered around other stars, how could we detect it? On Earth, there have been three distinct stages of progress to advanced life: (1) microbial anaerobes, (2) oxygen in the atmosphere, and (3) intelligence and technology (radio transmissions). These stages are shown in Figure 6.7 along with signatures for detection of signs of life. If a stage of advanced life is characterized by broadcasting radio waves, detection may be fairly easy. This is the principle behind most approaches to the Search for Extraterrestrial Intelligence (**SETI**) programs. The detection of intelligent radio emissions completely obviates the search for microbial life, liquid water, and planets within another star's habitable zone. The search is replaced with a dialog. Unfortunately, the duration of technological civilization on Earth has been infinitesimal compared with that of life overall. Furthermore, intelligence may be rare (Hart, 1975; McKay, 1996). This suggests that other search strategies need to be conducted in parallel with SETI.

The presence of oxygen in the atmosphere of an exoplanet can be detected most readily by searching for the spectral signature of ozone as the planet transits its central star (Owen, 1980; McKay, 2014). Taking Earth as an example, a search for oxygen or ozone is a strategy that would have yielded good results over only about half of Earth's biological history. To achieve a more inclusive result, one must search for the presence of liquid water (which on Earth coincided with biology). A search for liquid water must have two components. First, water vapor must be present in the atmosphere. This, however, is not enough, because a search of our Solar System would show water in the atmospheres of Venus, Earth, and Mars. Detection of atmospheric water vapor would have to be complemented by the measurement of the planetary surface temperature. This could be done, despite the presence of obscuring haze and clouds, by a variety of spectroscopic techniques suitable for transit and reflectance spectroscopy of exoplanets (Irwin *et al.*, 2014). For Venus, Earth, and Mars, the results would be 440°C, 15°C, and –60°C, respectively. The sole habitability of Earth for life as we know it among these three worlds would be established.

Complex life and Technology in the Universe

The ultimate search for life in the Universe is that for other technological civilizations. On Earth, life transitioned from microbial life to complex multicellular life and to a telescope-making species.

The evolution of advanced life must require more exacting conditions than for microbial life and could be closely tied to the planetary and stellar neighborhoods. One main hazard appears to be environmentally damaging impacts. The frequency of such catastrophes may need to be low enough to allow life to develop, but perhaps frequent enough to promote occasional rearrangements in the pattern of evolution. Climate stability is another factor in the development of advanced life. Certain levels of climate change may promote evolution and force the development of coping strategies that lead ultimately to technologically capable intelligence. Too much variability, however, may destroy advanced life forms. An important control on climate is **obliquity**, the tilt of a planet's spin axis with respect to its orbit around a star. Earth's substantial obliquity (23°) causes the seasons. For Earth, changes in obliquity are moderated by the presence of the Moon. Without a large close Moon, huge obliquity changes might occur; obliquity would vary chaotically (Laskar *et al.*, 1993). Astronomical effects can also endanger complex life; for example, ionizing radiation from nearby supernovae, cosmic ray bursts, and other sources of radiation (Lineweaver *et al.*, 2004). Microorganisms are at much less risk from these effects; complex life is much easier to destroy than microbial life (Cockell, 2003). These considerations suggest that complex life may be less common in the Universe than microbial lifeforms (Brownlee and Ward, 1999).

As discussed above, the rise of complex life on Earth appears to be tied to the rise of oxygen. The long interval of time between the first evidence of microbial life on Earth, and the appearance of multicellular organisms can be understood in terms of the size and recycling properties of the Earth. McKay (1996) suggested that smaller planets, such as Mars, could undergo a similar rise of oxygen and the associated biological developments hundreds, or thousands of times faster than in Earth.

It is often assumed that the rise of life occurred at an optimal stage in the evolution of the Universe — the ~10-billion-year interval between the Big Bang and the earliest evidence for life on Earth was necessary for the formation of the heavy elements. It is also assumed that the 3.5-billion-year interval between the appearance of life on Earth and telescope-making intelligence is a constant of the Universe. For example, Turnbull and Tarter (2003) developed a target selection for SETI and assumed that only stars 3 billion years old could have planets with intelligent life. Similarly, Norris (2000) assumes that technology only arises 5 billion years after the formation of the star.

Both the assumption of time in the Universe required for life and time on Earth required for technology should be questioned.

The formation of planets suitable for life has been assumed to be related to **metallicity** both in our galaxy (Gonzalez *et al.*, 2001 and see discussion in Viriginia Trimble's Chapter 3) and in other galaxies (Lineweaver, 2001; Suthar and McKay, 2012). The data from the Kepler mission provided the first direct evidence of the range of stellar metallicity for which rocky planets are present. Buchhave *et al.* (2012) conclude that the average metallicity of the stars with small planets is close to the value for the Sun. A listing of the seven most Earth-like planets discovered by Kepler (Schuler *et al.*, 2015) plus the reported Earth-like planets Kepler-186f (Quintana *et al.*, 2014) and Kepler-452 give an average metallicity very close to the Sun's, with a range from half to 15 times the Sun's value. This is in very good agreement with a pre-Kepler estimate of Lineweaver (2001) which suggested that the 68% confidence range in metallicity for Earth-like planet formation was in this same range. Based on estimates of the star-formation rate and of the gradual buildup of metals in the Universe, Lineweaver (2001) estimates that most Earth-like planets in the Universe are older than the Earth and that their average age is 1.8 ± 0.9 billion years older than the Earth.

The Origin and Evolution of Life in the Universe

Studies of the very early history of the Universe suggest that very massive (>100 solar masses) short-lived stars in the first generation of stars could have been significant sources of metallicity (Ohkubo et al., 2009; Aoki et al., 2014). Loeb (2014) suggests the possibility of habitable planets very early after the Big Bang, and Bialy et al. (2015) suggest that high abundances of water vapor could have existed in extremely low metallicity (10^{-3} solar) gas during the epoch of first metal enrichment of the interstellar medium at very early times in the Universe (< 1 billion years after the Big Bang) if that gas was shielded by H_2. Thus, it is possible that some planets appeared in favorable locations in the early Universe soon after its origin.

Carter (1983, 2008) proposed a mathematical approach to estimating the probability of technology on Earth-like planets. He assumed that the record of life and technology on Earth was a representative sample of the age and occurrence distribution of technology on Earth-like planets. In that perspective, the fact that technology arose on Earth not long before the planet becomes uninhabitable is an important constraint. This can be seen in Figure 6.3 which shows both the origin of technology and the expected end of habitability of the Earth due to the warming Sun. The interval is less than a half billion years (Lovelock and Whitefield, 1982; Caldeira and Kasting, 1992). Based on this comparably short interval between the rise of technology and the end of habitability, Carter (2008) suggested that there are at least five hard steps in the evolution of intelligence associated with advanced technology — it is not a simple, easily evolvable feature.

McKay (1996) suggested that from a biological perspective, distinct from a mathematical approach, the development of human-like species can be reduced to three essential steps: (1) the origin of life, which appears to have been a rapid, biochemical event; (2) the rise of oxygen and multicellularity, a slow, geologically determined event related to the size of the Earth; and (3) the development of intelligence capable of radio telescopes, an event with unknown causes, possibly random, possibly rare.

On Earth, the evolutionary path from complex life to technology did not take 3.5 billion years but it was not a straight one; there were considerable periods of time during which intelligence at the level of humans could have evolved but did not. Perhaps the most notable of such periods was the dinosaur era. If these creatures were as biochemically and behaviorally sophisticated as is now thought, the fact that they did not develop an intelligent species during the 150-million-year period in which they dominated Earth is puzzling (McKay, 1996). It has been suggested that dinosaur species that could have been on the road to technology existed as long as 12 million years before the demise of dinosaurs 65 million years ago (Russell and Séguin, 1982). In this context, it is interesting to note that humans developed from simple primates in less than 3 million years.

Thus, the record on Earth suggests that intelligent life can arise very rapidly, millions of years or less, on a small planet with lucky evolutionary events (McKay, 1996), but the record also indicates that the path is apparently random and not deterministic and thus could be so rare as to be consistent with human-like intelligence on Earth being singular in the history of the Universe (McKay, 1996). But rare may be just relative. By the zero-one-infinity rule mentioned before, if we are not the only intelligent species in the history of the Universe then there are likely to be innumerable such species.

Its life, Jim, but not as we know it.

Throughout this chapter, we have considered life only as we find it on Earth: water-based, carbon-built life. We have assumed that the essential features of life would be the same on another planet on which the main energy source was sunlight, the primary liquid was water, and the most naturally reactive element was carbon, found primarily in the form of CO_2. We have only one example of life; thus, it is certainly premature to deduce that any characteristics of this life are the absolute requirements for life elsewhere. On the other hand, a fruitful search strategy must be based on some model of life. Because we have only one example, the safest strategy is to search in similar environments for life of a similar type. This is a purely practical consideration, and does not imply that life very different from Earth-life cannot exist.

Two popular suggestions for alternative biochemical systems are the substitution of ammonia for water as the solvent for life's molecules, and the substitution of silicon for carbon as the basic structural element — although silicon atoms rarely make double bonds. Neither suggestion can be disproved at the present time, but neither can form the basis for a search strategy that is as well developed as the search for water–carbon life. We must first seek our cousins, and from there a broader understanding that can allow us to see strangers.

References

Abramov, O., and Mojzsis, S. J. 2009. Microbial habitability of the Hadean Earth during the late heavy bombardment. *Nature* 459(7245): 419–422.

Allwood, A. C., Walter, M. R., Kamber, B. S., Marshall, C. P., and Burch, I. W. 2006. Stromatolite reef from the early Archaean era of Australia. *Nature* 441(7094): 714–718.

Aoki, W., Tominaga, N., Beers, T. C., Honda, S., and Lee, Y. S. 2014. A chemical signature of first-generation very massive stars. *Science* 345(6199): 912–915.

Bialy, S., Sternberg, A., and Loeb, A. 2015. Water formation during the epoch of first metal enrichment. *The Astrophysical Journal Letters* 804(2): L29–33.

Borucki, W. J., Agol, E., Fressin, F., Kaltenegger, L., Rowe, J., Isaacson, H., and Fabrycky, D. 2013. Kepler-62: A five-planet system with planets of 1.4 and 1.6 Earth radii in the habitable zone. *Science* 340(6132): 587–590.

Brownlee, D., and Ward, P. 1999. *Rare Earth* (New York: Copernicus, Springer-Verlag).

Buchhave, L. A., Latham, D. W., Johansen, A., Bizzarro, M., Torres, G., Rowe, J. F., and Bryson, S. T. 2012. An abundance of small exoplanets around stars with a wide range of metallicities. *Nature* 486(7403): 375–377.

Cairns-Smith, A.G. 1982. *Genetic Takeover and the Mineral Origins of Life* (Cambridge, UK: Cambridge University Press).

Caldeira, K., and Kasting, J. F. 1992. The life span of the biosphere revisited. *Nature* 360(6406): 721–723.

Carter, B. 1983. The anthropic principle and its implications for biological evolution. *Philosophical Transactions of the Roy Society*. A310: 347–363.

Carter, B. 2008. Five-or six-step scenario for evolution? *International Journal of Astrobiology* 7(02): 177–182.

Catling, D. C., Glein, C. R., Zahnle, K. J., and McKay, C. P. 2005. Why O2 is required by complex life on habitable planets and the concept of planetary "oxygenation time". *Astrobiology* 5(3): 415–438.

Cavalier-Smith, T. 2010. Deep phylogeny, ancestral groups and the four ages of life. *Philosophical Transactions of the Royal Society of London B: Biological Sciences* 365(1537): 111–132.

Cockell, C. 2003. *Impossible Extinction: Natural Catastrophes and the Supremacy of the Microbial World* (Cambridge University Press).

Davies, R. E., and Koch, R. H. 1991. All the observed universe has contributed to life. *Philosophical Transactions of the Royal Society of London B: Biological Sciences* 334(1271): 391–403.

Davis, W. L., and McKay, C. P. 1996. Origins of life: A comparison of theories and application to Mars. *Origins of Life and Evolution of Biospheres* 26(1): 61–73.

Ehrenfreund, P., and Charnley, S. B. 2000. Organic molecules in the interstellar medium, comets, and meteorites: A voyage from dark clouds to the early Earth. *Annual Review of Astronomy and Astrophysics* 38(1): 427–483.

Eigen, M., Lindemann, B. F., Tietze, M., Winkler-Oswatitsch, R., Dress, A., and Von Haeseler, A. 1989. How old is the genetic code? Statistical geometry of tRNA provides an answer. *Science* 244(4905): 673–679.

Freissinet, C., Glavin, D. P., Mahaffy, P. R., Miller, K. E., Eigenbrode, J. L., Summons, R. E., and Franz, H. B. 2015. Organic molecules in the Sheepbed Mudstone, Gale Crater, Mars. *Journal of Geophysical Research: Planets* 120(3): 495–514.

Gilbert, W. 1986. Origin of life: The RNA world. *Nature* 319(6055): 618.

Gonzalez, G., Brownlee, D., and Ward, P. 2001. The galactic habitable zone: Galactic chemical evolution. *Icarus* 152(1): 185–200.

Grotzinger, J. P., Gupta, S., Malin, M. C., Rubin, D. M., Schieber, J., Siebach, K., Sumner, D. Y. et al. 2015. Deposition, exhumation, and paleoclimate of an ancient lake deposit, Gale crater, Mars. *Science* 350 (6257): aac7575, DOI: 10.1126/science.aac7575.

Hart, M. H. 1975. Explanation for the absence of extraterrestrials on Earth. *Quarterly Journal of the Royal Astronomical Society* 16: 128–135.

Hecht, M. H., Kounaves, S. P., Quinn, R. C., West, S. J., Young, S. M. M., Ming, D. W., and DeFlores, L. P. 2009. Detection of perchlorate and the soluble chemistry of martian soil at the Phoenix lander site. *Science* 325(5936): 64–67.

Irwin, P. G., Barstow, J. K., Bowles, N. E., Fletcher, L. N., Aigrain, S., and Lee, J. M. 2014. The transit spectra of Earth and Jupiter. *Icarus* 242: 172–187.

Jenkins, J. M., Twicken, J. D., Batalha, N. M., Caldwell, D. A., Cochran, W. D., Endl, M., and Petigura, E. 2015. Discovery and validation of Kepler-452b: A 1.6 R super Earth exoplanet in the habitable zone of a G2 star. *The Astronomical Journal* 150(2): 56–74.

Knoll, A. H. 1985. The Precambrian evolution of terrestrial life. In Papagiannis, M. D. (Ed.) *The Search for Extraterrestrial Life: Recent Developments* Netherlands: Springer, pp. 201–211.

Kushner, D. 1981. Extreme environments: Are there any limits to life? In *Comets and the Origin of Life* Netherlands: Springer, pp. 241–248.

Laskar, J., Joutel, F., and Robutel, P. 1993. The stabilization of the Earth's obliquity by the Moon. *Nature* 361: 615–617.

Lazcano, A., and Miller, S. L. 1994. How long did it take for life to begin and evolve to cyanobacteria? *Journal of Molecular Evolution* 39: 546–554.

Lehninger, A. L. 1975, *Biochemistry* (New York: Worth).

Lineweaver, C. H. 2001. An estimate of the age distribution of terrestrial planets in the universe: Quantifying metallicity as a selection effect. *Icarus* 151(2): 307–313.

Lineweaver, C. H., Fenner, Y., and Gibson, B. K. 2004. The galactic habitable zone and the age distribution of complex life in the Milky Way. *Science* 303(5654): 59–62.

Loeb, A. 2014. The habitable epoch of the early Universe. *International Journal of Astrobiology* 13(04): 337–339.

Lovelock, J. E., and Whitfield, M. 1982. Life span of the biosphere. *Nature* 296: 561–563.

Lyons, T. W., Reinhard, C. T., and Planavsky, N. J. 2014. The rise of oxygen in Earth/'s early ocean and atmosphere. *Nature* 506(7488): 307–315.

McCollom, T. M. 2013. Miller-Urey and beyond: What have we learned about prebiotic organic synthesis reactions in the past 60 years? *Annual Review of Earth and Planetary Sciences* 41: 207–229.

McKay, C. P. 1986. Exobiology and future Mars missions: The search for Mars' earliest biosphere. *Advances in Space Research* 6(12): 269–285.

McKay, C. P. 1991. Urey prize lecture: Planetary evolution and the origin of life. *Icarus* 91(1): 93–100.

McKay, C. P. 1996. Time for intelligence on other planets. In Doyle, L. R. (Ed.), *Circumstellar Habitable Zones*, Vol. 1, (Menlo Park: Travis House), p. 405–419.

McKay, C. P. 2001. The search for a second genesis of life in our Solar System. In *First Steps in the Origin of Life in the Universe* Netherlands: Springer, pp. 269–277.

McKay, C. P. 2004. What is life-and how do we search for it in other worlds? *PLoS Biology* 2: 1260–1262.

McKay, C. P. 2014. Requirements and limits for life in the context of exoplanets. *Proceedings of the National Academy of Sciences* 111(35): 12628–12633.

McKay, C. P., Anbar, A. D., Porco, C., and Tsou, P. 2014. Follow the plume: The habitability of Enceladus. *Astrobiology* 14(4): 352–355.

McKay, C. P., and Davis, W. L. 1991. Duration of liquid water habitats on early Mars. *Icarus* 90: 214–221.

McKay, C. P., and Stoker, C. R. 1989. The early environment and its evolution on Mars: implication for life. *Reviews of Geophysics* 27(2): 189–214.

Melosh, H. J. 1988. The rocky road to panspermia. *Nature* 332(6166): 687–688.

Miller, S. L. 1953. A production of amino acids under possible primitive earth conditions. *Science* 117(3046): 528–529.

Navarro-González, R., Vargas, E., de La Rosa, J., Raga, A. C., and McKay, C. P. 2010. Reanalysis of the Viking results suggests perchlorate and organics at midlatitudes on Mars. *Journal of Geophysical Research: Planets (1991–2012)* 115(E12).

Norris, R. P. 2000. How old is ET? *Acta Astronautica* 47(2): 731–733.

Ohkubo, T., Nomoto, K. I., Umeda, H., Yoshida, N., and Tsuruta, S. 2009. Evolution of very massive Population III stars with mass accretion from pre-main sequence to collapse. *The Astrophysical Journal* 706(2): 1184–1208.

Orgel L. E. (1998) The origin of life — how long did it take? *Orig. Life. Evol. Biosph.*, 28, 91–96.

Owen, T. 1980. The search for early forms of life in other planetary systems: Future possibilities afforded by spectroscopic techniques. In *Strategies for the Search for Life in the Universe* Netherlands: Springer, pp. 177–185.

Pizzarello, S. 2004. Chemical evolution and meteorites: An update. *Origins of Life and Evolution of the Biosphere* 34(1–2): 25–34.

Quinn, R. C., Martucci, H. F., Miller, S. R., Bryson, C. E., Grunthaner, F. J., and Grunthaner, P. J. 2013. Perchlorate radiolysis on Mars and the origin of martian soil reactivity. *Astrobiology* 13(6): 515–520.

Quintana, E. V., Barclay, T., Raymond, S. N., Rowe, J. F., Bolmont, E., Caldwell, D. A., and Lissauer, J. J. 2014. An Earth-sized planet in the habitable zone of a cool star. *Science* 344(6181): 277–280.

Robertson, M. P., and Joyce, G. F. 2012. The origins of the RNA world. *Cold Spring Harbor Perspectives in Biology* 4(5): a003608.

Russell, D. A., and Séguin, R. 1982. Reconstructions of the small Cretaceous theropod, Stenonychosaurus inequalis, and a hypothetical dinosauroid. Syllogeus Number 37, National Museum of Natural Sciences, Ottawa, Canada.

Schidlowski, M. 1988. A 3,800-million-year isotopic record of life from carbon in sedimentary rocks. *Nature* 333(6171): 313–318.

Schuler, S. C., Vaz, Z. A., Santrich, O. J. K., Cunha, K., Smith, V. V., King, J. R., and Isaacson, H. 2015. Detailed abundances of stars with small planets discovered by Kepler. I. The first sample. *The Astrophysical Journal* 815(1): 5–26.

Suthar, F., and McKay, C. P. 2012. The galactic habitable zone in elliptical galaxies. *International Journal of Astrobiology* 11(03): 157–161.

Tice, M. M., and Lowe, D. R. 2004. Photosynthetic microbial mats in the 3,416-Myr-old ocean. *Nature* 431(7008): 549–552.

Turnbull, M. C., and Tarter, J. C. 2003. Target selection for SETI. I. A catalog of nearby habitable stellar systems. *The Astrophysical Journal Supplement Series* 145(1): 181–198.

Waite Jr, J. H., Lewis, W. S., Magee, B. A., Lunine, J. I., McKinnon, W. B., Glein, C. R., and Nguyen, M. J. 2009. Liquid water on Enceladus from observations of ammonia and 40Ar in the plume. *Nature* 460(7254): 487–490.

Woese, C. R. 1987. Bacterial evolution. *Microbiological Reviews* 51(2): 221–271.

Zuckerman, B., and Young, E.D., 2018. Characterizing the chemistry of planetary materials around white dwarf stars. In Deeg, H. and Belmonte, J. (Eds.) *Handbook of Exoplanets* (Switzerland: Spinger).

Glossary

Abiotic synthesis. Production of organic molecules by nonbiological processes (i.e., chemical reactions in the absence of living systems).

Absorption line. A wavelength (or small range of wavelengths) at which the brightness of a spectrum is lower than it is at neighboring wavelengths.

Accretion. The transfer of matter to the surface of a star, often from a companion star in a binary system. When the transferred matter goes into orbit around a star or black hole, an accretion disk is formed.

Active galaxy. A galaxy whose nucleus emits large quantities of electromagnetic radiation that does not appear to be produced by stars.

Adjustable parameter. In an equation a constant that can take on a variety of numerical values (sometimes including zero) so that the equation yields a family of related solutions, not just one solution.

Ambipolar diffusion. Process in which magnetic fields slowly drift outwards through a plasma.

Amino acids. Nitrogen-containing acids, some of which make up the building blocks of proteins.

Ångstrom (Å). A unit of length commonly used for visible wavelengths of light; 1 Å = 10^{-8} cm.

Angular momentum. A measure of the amount of spin of an object; dependent on the object's rotation rate, mass, and mass distribution.

Antiparticle. A particle whose charge (if not neutral) and certain other properties are opposite those of a corresponding particle of the same mass. An encounter between a particle and its antiparticle results in mutual annihilation and the production of high-energy photons.

Astronomical unit (AU). Average distance between Earth and the Sun (150,000,000 km).

Baryonic matter. Protons, neutrons, and electrons, the components of ordinary matter.

Big Bang. The birth of the Universe in a very hot, dense state about 14 billion years ago, followed by the expansion of space.

Binary pulsar. A pulsar in a binary system. Often this term is used for systems in which the pulsar's companion is another neutron star.

Binary X-ray sources. Binary stars in which a white dwarf, a neutron star, or a black hole accretes matter from a normal companion, which produces large quantities of X-rays.

Binding energy. The difference in mass, expressed in energy units, when two or more particles are closely bound to each other compared with when they are far apart.

Bipolar outflow. A strong, well-collimated outflow of material from a young stellar object. The outflow is narrow in angular extent and is generally confined to the rotational poles of the system.

Blackbody. An object with a constant temperature that absorbs all radiation that hits it.

Black hole. A region of space-time in which the gravitational field is so strong that nothing, not even light, can escape. Predicted by Einstein's general theory of relativity.

Centrifugal force. The outward force felt by an object in a rotating frame of reference.

Cepheid. A type of pulsating variable star with a luminosity that can be determined from the period of its variation. Cepheids with long pulsation periods are bigger and thus more luminous than short-period Cepheids.

Chandrasekhar mass limit. Maximum stable mass for a degenerate star (white dwarf) or degenerate iron core in a massive star. Configurations more massive than the Chandrasekhar mass — about 1.4 solar masses, depending on composition — collapse or explode. Also called Chandrasekhar limit.

Charge-coupled device (CCD). A solid-state imaging chip whose properties include high sensitivity, large dynamic range, and linearity.

Chemical evolution. 1) Gradual buildup of the heavy element supply in a galaxy and accompanied by changes in the star formation rate, color, and luminosity of the galaxy; 2) Complex of chemical reactions that must occur on the surface of Earth or other inhabited planet before self-replicating molecules and biological evolution can appear.

Chemical reactions. Reactions that assemble or destroy chemical compounds through lending or sharing of electrons among atoms.

Cherenkov radiation. Light produced by charged particles moving through a transparent medium with a speed exceeding the local speed of light.

Circumstellar disk. A nebula of gas and dust orbiting a star. The circumstellar disks produced by the star formation process typically have radial sizes comparable to that of our solar system.

CNO tri-cycle. Linked cycles of nuclear reactions that transform hydrogen to helium, with carbon, nitrogen, and oxygen nuclei acting as catalysts.

Cold dark matter. A type of dark matter that was moving at much less than the speed of light 10,000 years after the Big Bang (see "**Dark matter**").

Column density. The total mass per unit area along a given line of sight through an astrophysical object such as a molecular cloud, a protostellar envelope, or an accretion disk. Proportional to the number of molecules (or dust grains) per unit area along a given line of sight.

Continuum: The radiation emitted by an astrophysical object can be separated into two parts. The continuum is the base-level of the radiation and covers a wide range of wavelengths. Superimposed on this continuum radiation are narrow spikes of radiation, called lines, that either add or subtract from the continuum level.

Convection. A process that transports heat (energy) by motions of the fluid itself.

Core of a main-sequence star. The central region where the temperature and pressure conditions enable energy production through thermonuclear fusion of hydrogen into helium. In an evolved star, usually refers to the degenerate central region.

Cosmic microwave background. Diffuse glow of light, in the microwave (radio) part of the spectrum, left from the cooling Big Bang at an age of a few hundred thousand years.

Cosmic rays. Very high-energy protons and other particles found throughout interstellar space; their impacts on Earth's atmosphere make carbon-14 and cause mutations.

Cosmological constant. In Einstein's general relativity equations, a term that leads to an acceleration of the expansion of the Universe.

Cosmology. The study of the overall structure and evolution of the Universe.

Critical density. Density the Universe would have if its expansion rate were just barely sufficient to prevent a recollapse.

Dark Energy. A theoretical respulsive force that counteracts gravity and causes the universe to exapand at an accelerating rate.

Dark matter. Apparently unseen matter that dominates the mass of the Universe. Many astronomers believe dark matter is made of a gas of exotic, weakly interacting particles. According to this theory, embryonic fluctuations in the matter density of the particles grew into galaxies, galaxy clusters, and galaxy superclusters.

Degenerate gas. Gas that is so dense its electrons can no longer move freely through space; the gas pressure depends only on the gas density, not on its temperature.

Density fluctuation. A region in which the amount of matter per unit volume is slightly higher or lower than average. Somewhat analogous to a high or low pressure zone in Earth's atmosphere.

Deuterium. Heavy hydrogen (with a nucleus consisting of one proton and one neutron). See **isotope**.

Dipole. A pattern where one side of the sky is hot and the opposite side is cold.

Dipole field. The pattern of electric field lines produced by a pair of equal and opposite electric charges, or of magnetic field lines surrounding a bar magnet.

DNA (Deoxyribonucleic acid). Nucleic acid with the structure of a double helix that serves as the genetic memory for life.

Doppler. Nineteenth century physicist who discovered the variation in the wavelength of waves caused by the motion of the source of the waves relative to the observer.

Doppler shift. The change in wavelength or frequency produced when a source of waves and the observer move relative to each other. Blueshifts (to shorter wavelengths) and redshifts (to longer wavelengths) are associated with approach and recession, respectively.

$E = mc^2$. Einstein's famous formula for the equivalence of energy and mass.

Eccentricity. A measure of the deviation of an elliptical orbit from a circle.

Electromagnetic force. One of the four forces of nature. Electromagnetic interactions hold electrons in atoms, hold atoms in molecules, and are used in all electronic devices.

Electromagnetic radiation. Self-propagating, oscillating, electric and magnetic fields; from shortest to longest wavelengths, the categories include gamma rays, X-rays, ultraviolet, optical (visible), infrared, and radio.

Electron. Negatively charged, low-mass particle that orbits atomic nuclei in atoms; normally the number of electrons is the same as the number of protons.

Electron degeneracy. A peculiar state of matter at high densities in which, according to the laws of quantum mechanics, electrons move very rapidly in well-defined energy levels and exert tremendous pressure on their surroundings.

Electron volt (eV). Energy necessary to raise an electron through a potential difference of one volt. One watt for 1 second is equivalent to 6 billion-billion eVs. Using the conversion $E = mc^2$, the rest mass of an electron is about 511,000 eV.

Electroweak. A unified force that combines the electromagnetic and weak nuclear interactions. Predicted by Weinberg and Salam and experimentally verified by Rubbia and van der Meer.

Elliptical galaxy. A system of 1 billion to 100 billion stars bound together by gravity. Ellipticals have basically round shapes, but typically have one long axis and one short axis, which may differ by a factor of 2; their true shapes are either like onions, footballs, or both. Unlike spiral galaxies, they have no thin disks of stars, gas, and dust and appear to have formed few new stars in the recent past. The stars in an elliptical have a wide variety of orbits in various directions, which leads to an overall appearance of random motion.

Emission line. A wavelength (or small range of wavelengths) at which the brightness of a spectrum is higher than it is at neighboring wavelengths.

Endosymbionts: Organisms living inside a host organism in a symbiotic way.

Entropy: A measure of the amount of disorder in a physical system, formally defined as the logarithm of the number of accessible states. Systems that are more disordered have a higher entropy.

Erg. A unit of energy (work) in the metric system equal to a force of one dyne (g-cm/s^2) acting through a distance of 1 cm; 10^7 ergs = 1 Joule.

Escape velocity. The minimum speed which an object must have to escape the gravitational pull of another object.

Event horizon. Boundary of a black hole from within which nothing can escape (*see* **Schwarzschild radius**).

Exoplanet. A planet which orbits a star other than our Sun.

Expansion of the Universe. The observed characteristic of the Universe that galaxies are all receding from each other with speeds that are in direct proportion to their separations (see **Hubble's Law**). Thus, a galaxy twice as far from Earth as another galaxy will have a recession velocity twice as great as the closer galaxy. Galaxies do not actually move through space; space itself is expanding and thus carrying galaxies along like sequins painted on an expanding balloon.

Fossil. Traces and remains of past life on Earth.

Fractal. A pattern having the mathematical property that any small part of it, viewed at any magnification, shares the same statistical character as the original, and is thus indistinguishable from the whole. Such patterns are common in nature (e.g., a coastline).

Fusion. Nuclear reaction in which two or more light nuclei combine to form a heavier one; some fusion reactions in stars are also called nuclear burning or simply burning.

Galaxy. A large, gravitationally bound collection of stars such as the Milky Way galaxy, which consists of several hundred billion stars. Galaxy shapes are generally spiral or elliptical, and sometimes irregular or peculiar.

Gamma ray. A very high-energy photon, more energetic than an X-ray.

Gas chromatograph mass spectrometer. An instrument that separates gases in a mixture and then determines the molecular mass of each component so that the component can be identified.

G dwarf problem. Near the Sun, the unexpected paucity of long-lived stars with less than about 10% of the solar metallicity.

General theory of relativity. Albert Einstein's comprehensive theory of mass, space, and time. According to the theory, the gravitational field associated with matter produces a curvature of space-time in its vicinity.

Grand unified theory (GUT). A model for unifying the strong nuclear force, the weak nuclear force, and the electromagnetic force into a single interaction. Several GUTs have been proposed, but not experimentally verified.

Grating. A piece of glass having many parallel grooves cut into its surface. Used to disperse light into its component wavelengths.

Gravitational waves. According to relativity theory, waves ("ripples" in space-time) emitted because of changes in the distribution of matter.

Great Attractor. A proposed nearby supercluster of galaxies in a region of the sky largely obscured by the Milky Way, and thus difficult to study.

Hα. Photon produced by a hydrogen atom when its electron jumps down from the third to the second energy level.

Habitable planets. Planets with liquid water, a supply of elements needed for biochemistry, and an environment stable for billions of years.

Glossary

Half-life. The time it takes for half a given quantity of a radioactive substance to decay.

Heavy bombardment. Rain of asteroidal and comet-like material (**planetesimals**) that comprised the final formation of the planets.

Homogeneous. The same at all locations (e.g., homogenized milk is not separated into cream and milk).

Horizon. Edge of the *visible* Universe, but not the actual edge of the Universe (because the Universe has no edge).

Hot Big Bang. A model of the Universe beginning at very high density and temperature, which expands and cools to become like the Universe we observe now.

Hot dark matter. A type of dark matter that was moving at a substantial fraction of the speed of light 10,000 years after the Big Bang (see "**Dark matter**").

Hubble constant. The observed slope of the line defined by **Hubble's Law**, it has the dimensions of inverse time. Thus the inverse of the Hubble Constant is often referred to as the "Hubble Time".

Hubble's Law. The linear correlation of a galaxy's recession velocity with its distance from Earth. It is a natural observational consequence of our uniformly expanding Universe (see **Expansion of Universe**).

Hydrothermal vents. Fissures, usually on the ocean floor, through which hot water and reactive gases emanate.

Inflationary scenario. A modification of the Big Bang model in which a large cosmological constant exists temporarily early in the history of the Big Bang and leads to a rapid accelerating expansion of the Universe, which is then followed by the normal Big Bang model of the expansion.

Initial mass function (IMF). Distribution of stellar masses of a given stellar population at its moment of birth; the IMF is not the same as the distribution of stellar masses seen later because stars with different masses have different lifetimes.

Interstellar medium. Space between the stars; filled to some extent with gas and dust.

Ionized. Having lost at least one electron. Atoms become ionized primarily by the absorption of energetic photons and by collisions with other particles.

Isothermal. An object is isothermal if all of its parts have the same temperature. This condition arises when the cooling processes are efficient, so that heat is removed and the system reaches a uniform (low) background temperature. The dense inner regions of clouds that collapse to form stars have this property.

Isothermal equation of state. A relation between the pressure P, the density ρ, and the temperature T of a gas. When the temperature of a gas remains the same, an isothermal equation of state has the form $P = a^2\rho$, where the sound speed a is a constant.

Isotopes. Atomic nuclei having the same number of protons (and, hence, almost identical chemical properties) but different numbers of neutrons and, therefore, somewhat different masses.

Isotropic. The same in all directions.

Kelvin (K). The size of 1 degree on the Kelvin temperature scale, in which absolute zero is 0 K, water freezes at 273 K, and water boils at 373 K. To convert from the Kelvin scale to the Celsius (centigrade; C) scale, subtract 273 from the Kelvin scale value.

Keplerian orbits. Orbits that obey the classical laws of Kepler.

Kepler's third law. If one object orbits another, the square of its period of revolution is proportional to the cube of the major axis of the elliptical orbit.

Kilonova. An energetic transient event that occurs in a compact binary system, when two neutron stars, or a neutron star and a black hole, merge with each other.

Large Magellanic Cloud (LMC). A dwarf companion galaxy of our Milky Way galaxy, located about 170,000 light years away; best seen from Earth's southern hemisphere.

Laser Interferometer Gravitational-wave Observatory (LIGO). A multi-facility observatory for detection of minute spatial distortions caused by passing gravitational waves.

Lichen. A life-form based on the symbiotic association of fungus and algae.

Light curve. A plot of an object's brightness as a function of time.

Lightyear. Distance light travels through a vacuum in 1 year; about 10 trillion km.

Luminosity. Energy per unit of time generated by an astrophysical object.

Magnetar. A neutron star with an ultra-strong magnetic field about 10^{15} Gauss.

Magnitude. A scale used by astronomers to measure apparent brightness. Each 5 units on the magnitude scale corresponds to a 100-fold decrease in the energy flux. The Sun has magnitude –26.5. Sirius, the brightest star in the night sky, has magnitude −1.6. The faintest stars visible with the naked eye have magnitude 6.

Main-sequence. The phase of stellar evolution, lasting about 90% of a star's life, during which the star fuses hydrogen to helium in its core.

Metallicity. The fraction by mass of a star, galaxy, or gas cloud that is made up of all elements heavier than helium (not just actual metals).

Millisecond pulsar. A pulsar whose period is roughly in the range 1 to 20 milliseconds.

Molecular cloud. A large cloud of interstellar gas in the molecular state. These clouds form stars but are much larger than a single star; the clouds typically have masses between 10^4 and 10^6 solar masses.

Molecular Cloud Core: A dense region within a gas cloud that provides the birth place for an individual star. These cores contain more mass than the stars they form, but are much smaller and denser than the clouds.

N-body (computer) simulation. Following the growth of structure through a computer program that calculates the gravitational force between N bodies (where N is a large number) representing the total number of objects.

Nebula. A region containing an above-average density of interstellar gas and dust.

Neutralino. A particle predicted by supersymmetric models for the forces of nature. The models predict that each type of known particle will have a supersymmetric partner. The neutralino is the lightest, electrically neutral, supersymmetric partner and is a candidate for cold dark matter. As of 2020, no supersymmetric partner particles of any kind have been observed experimentally.

Neutrino. A very low-mass, or massless, uncharged particle that interacts exceedingly weakly with matter after being created by certain nuclear reactions. There are three types: electron, muon, and tau neutrinos.

Neutron. Uncharged, massive particle found in nuclei of atoms; different **isotopes** of a given element have different numbers of neutrons in their nuclei.

Neutron degeneracy. Similar to **electron degeneracy**, but with neutrons replacing electrons. Sets in at a much higher density and pressure than electron degeneracy.

Neutron star. The compact endpoint in stellar evolution in which 1 to 2 solar masses of material is compressed into a small (diameter = 20–30 km) sphere supported by **neutron degeneracy** pressure.

Nonbaryonic. Not made up of neutrons and protons, and thus not like any of the known chemical elements.

Nonlinear: Equations are nonlinear if the functions involved appear as variables of a polynomial of degree higher than one. In practice, equations become nonlinear when small fluctuations become large enough to cause a backreaction on the original system.

Nonthermal: Thermal motions result from the distribution of particle speeds for a gas interacting with a heat reservoir with a well-defined temperature. Nonthermal motions of the particles are those that are present in addition to the thermal motion.

Nova. Nonfatal stellar explosion caused by burning of degenerate hydrogen on the surface of a white dwarf in a binary system.

Nuclear fusion. Reactions in which low-mass atomic nuclei combine to form a more massive nucleus.

Nuclear reaction. A reaction that transforms one type of atomic nucleus into another by changing the number of neutrons and protons.

Nucleosynthesis. The creation of elements through nuclear reactions.

Obliquity. The angle between the spin axis of a planet and the direction perpendicular to its orbital plane. The larger the obliquity, the stronger are seasonal climate variations.

Opacity. The degree to which matter restricts the flow of electromagnetic radiation; a perfectly transparent substance has zero opacity.

Organic material. Compounds and molecules that contain carbon.

Oxidants. Compounds that cause binding with oxygen.

Oxidation-reduction reactions: Chemical reactions in which an electron is exchanged. In the oxidation-reduction reaction $2H_2 + O_2 = 2H_2O$, each H gives up one electron which is taken by the O.

Panspermia. The idea that life was carried to Earth from elsewhere.

Parsec. A unit of distance used by astronomers, equal to 3.09×10^{13} km (3.26 light-years).

Perchlorate: A salt of chlorine in which the chlorine atom is in its most oxidized state. Stable at room temperature.

Phase transition. A change between two different phases of matter, such as a solid melting and becoming a liquid, or a liquid boiling to become a gas.

Photon. A quantum, or package, of electromagnetic radiation that travels at the speed of light; from highest to lowest energies the categories include gamma rays, X-rays, ultraviolet, optical (visible), infrared, and radio.

Photosynthesis. Production of organic material by sunlight.

Phylogenetic tree. Relationship by evolutionary descent of a group of organisms or species.

Planetary nebula. A shell of gas, expelled by a red giant near the end of its life (but before the white dwarf stage), which glows because it is ionized by ultraviolet radiation from the star's remaining core.

Planetesimals. See **Heavy bombardment**.

Positron. Antiparticle of an electron.

Precession. The relatively slow movement of the axis of a spinning entity around another axis, usually due to a torque exerted on the system.

Pre-main-sequence star. A young star that is not generating energy exclusively through the fusion of hydrogen. Pre-main sequence stars initially generate energy through gravitational contraction and eventually evolve into stellar configurations in which hydrogen fusion takes place.

Primordial soup. Mixture of abiotically produced molecules, thought to have led to the origin of life.

Progenitor. In the case of a supernova, the star that will eventually explode.

Proteins. A class of biomolecules constructed from sequences of amino acids.

Proton. Positively charged, massive particle in nuclei of atoms; different chemical elements have different numbers of protons in their nuclei.

Proton-proton chain. Sequence of nuclear reactions that transforms hydrogen to helium.

Protostar. A star that is still in the process of forming. The star itself is generally deeply embedded within an infalling envelope of dust and gas.

Pulsar. An astronomical object that is detected through pulses of radiation (usually radio waves) having a short, extremely well-defined period; now thought to be a rotating neutron star with a very strong magnetic field.

Quantum fluctuations. The uncertainty principle in quantum mechanics leads to all allowed interactions occurring with some probability (see **vacuum energy density**).

Quantum mechanics. A twentieth-century theory that successfully describes the behavior of matter on very small scales (such as atoms) and radiation.

Quasar. The nucleus of an active galaxy. This nucleus is unusually luminous, perhaps 10 to 1000 times more powerful than the rest of the galaxy. Their ultimate energy source is suspected to be accretion onto a massive black hole. Most quasars are at distances of billions of light-years from Earth.

Radioactive nucleus. A nucleus capable of spontaneously emitting an electron, a helium nucleus, or a gamma ray.

Radiogenic heating. Heating caused by energy released by the decay of radioactive elements.

Red giant. Evolutionary phase following the main-sequence of a solar-type star; the star becomes both large and bright (though cool on the surface). Hydrogen burns in a shell around a helium core.

Redshift. Doppler shift for objects receding from Earth causes the wavelengths of light to get longer, and hence visible light shifts toward the red part of the spectrum. Lengthening of waves can also be caused by propagation in a strong gravitational field.

Reducing: The ability of an element or compound to lose (or "donate") an electron to another chemical species (the oxidizing species) in a chemical reaction.

Reducing atmosphere. An atmosphere that is rich in molecules that contain hydrogen, for example, ammonia (NH_3) and methane (CH_4).

Relative humidity. Ratio of actual vapor pressure of water to its saturation vapor pressure at a given atmospheric pressure and temperature.

Rest mass. The mass of an object that is at rest with respect to the observer. Massless particles, such as photons, *must* travel at the speed of light, but particles having nonzero rest mass *cannot* reach the speed of light.

RNA (Ribonucleic acid). Information-transporting molecule related to DNA; involved in protein synthesis.

Schwarzschild radius. 1) Radius to which a given mass must be compressed to form a nonrotating black hole. 2) Radius of the **event horizon** of a nonrotating black hole.

Sedimentary deposits. Material accumulated by water or wind.

SETI. The Search for Extraterrestrial Intelligence.

Shock wave. A compressional wave, characterized by a discontinuous change in pressure, produced by an object traveling through a medium faster than the local speed of sound.

Singularity. A mathematical point of zero volume associated with infinite values for physical parameters such as density.

Solar mass ($M_\odot$). The mass of the Sun, 1.99×10^{33} g, about 330,000 times the mass of Earth.

Space-time. The four-dimensional fabric of the Universe whose points are events having specific locations in space (three dimensions) and time (one dimension).

Glossary

Spallation. Breaking up or eroding of something fairly solid; in particular, the breakup of heavy atomic nuclei when hit by **cosmic rays**.

Specific Angular Momentum: Angular momentum per unit mass, where the angular momentum is the amount of momentum times the lever arm through which it acts. For orbital motion in a circle, the magnitude of the specific angular momentum is the speed times the radius of the circle.

Spectral energy distribution. A graph showing how much energy (in radiation) is emitted by an astronomical object as a function of the frequency (or wavelength) of the light.

Spectrograph. An instrument which displays the brightness of electromagnetic radiation from an object as a function of wavelength or frequency (see **Spectrum**).

Spectrum. A plot of the brightness of electromagnetic radiation from an object as a function of wavelength or frequency.

Spiral galaxy. A galaxy, like our Milky Way, made up of 1 billion to 100 billion stars bound together by gravity. The spiral pattern, found in the thin disks of stars, gas, and dust that surround a spheroidal bulge of stars, is highlighted by sites of continuing star birth. The stars in the bulge swarm in many directions, as in the closely related elliptical galaxies; most stars within the disk trace near-circular orbits around the center of the galaxy.

Steady State. A model of the expanding Universe with constant density and all other physical properties. Because of the expansion of the Universe, matter must be continually created to maintain constant density.

Stellar population. Mix of stars of different masses, with their various temperatures, metal abundances, and ages, that makes up a large system such as a star cluster or galaxy.

Stellar wind. Continuous or quasicontinuous release of gas from the outer atmosphere of a star.

Strong nuclear force. One of the four forces of nature. The strong nuclear force holds the particles in the nucleus of atoms together.

Supergiant. Evolutionary phase following the main-sequence of a massive star; the star becomes very bright and cool (red) or hot (blue); a sequence of nuclear reactions occurs in the stellar interior.

Supernova. Violent explosion of a star at the end of its life. Hydrogen is present or absent in the spectra of Type II or Type I supernovae, respectively.

Supernova remnant. Cloud of chemically enriched gases ejected into space by a supernova.

Tidal force. Difference between the gravitational force exerted by one body on the near and far sides of another body.

Time dilation. According to relativity theory, the slowing of time perceived by an observer watching another object moving rapidly or located in a strong gravitational field.

Transit. A temporary dimming of a star, when an exoplanet (or Venus or Mercury) passes between the star and the Earth.

T Tauri star. A newly formed (**pre-main-sequence**) star with many characteristic signatures of youth, including strong emission lines and excess infrared radiation. The name derives from the star T Tauri, which is the prototype for this class of very young stars.

Turbulence: Chaotic unsteady motions in a fluid.

Ultraviolet light. Light with wavelengths shorter than blue light, that is, wavelengths between 100 and 300 nm.

Vacuum energy density. Quantum theory requires empty space to be filled with particles and antiparticles being continually created and annihilated. This could lead to a net density of the vacuum, which if present, would behave like a cosmological constant (see **quantum fluctuations**).

Variable star. A star whose apparent brightness changes with time.

Virgo Supercluster. A large concentration of galaxies in the direction of the constellation Virgo, with a recession velocity of 10^3 km/s with respect to the Milky Way.

Weak nuclear force. One of the four forces of nature. The weak nuclear force is responsible for some radioactive decays as well as some of the fusion reactions in the Sun that provide heat and light for Earth.

White dwarf. Evolutionary end point of stars that have initial masses less than about 8 times the solar mass. All that remains is an electron degenerate core of helium, carbon-oxygen (the majority of cases), or oxygen-neon-magnesium.

Wormhole. Connection between two black holes in separate universes or in different parts of our Universe. Also called Einstein-Rosen bridge.

Index

A
abiotic synthesis, 195
Abramson, Louis, 53
accretion, 40, 90, 112, 113, 126, 132, 133, 141, 142, 150, 160, 164, 167, 168, 170, 173, 175, 176, 181
ALMA, 59, 60, 162, 163, 171
Alpher, Ralph, 76
amino acids, 190, 192
Andromeda nebula (Galaxy), 11, 26, 27, 34, 35, 39, 41, 42
angular momentum, 130, 138, 153, 158, 160, 166, 168, 170, 173, 175, 176
archaea, 192, 193
asteroids, 89, 107, 108, 177, 179, 199

B
Baade, Walter, 34, 35, 115, 126
bacteria, 64, 190–193, 198
Barish, Barry, 142
Bell, Jocelyn, 127, 134
Bethe, H., 78, 80
Big Bang, 1, 2, 6, 9, 11–16, 18–21, 23, 25, 28, 34, 36, 43, 44, 53, 55–57, 60, 63, 69, 74–77, 79, 81, 88, 94, 102, 124, 138, 141, 144, 206, 207
bipolar outflow, 153, 161, 167
blackbody, 10, 11, 168

black holes, 38–43, 55, 58, 59, 60, 87, 95, 99, 102, 108, 122, 128, 132–145, 182
Brahe, Tycho, 116
brown dwarfs, 165, 173–176, 180
Burbidge, G. and M., 70, 71, 74

C
carbon, 32–34, 60, 64, 66–68, 70, 77–85, 88, 89, 91, 100, 101, 112, 114, 194, 195, 198, 199, 208
centrifugal, 128, 158, 163, 164, 172
Cepheid, 27, 123, 127
Chadwick, James, 115
Chandrasekhar limit, 84, 112–114, 139
chemical evolution, 72, 90, 92–94, 165
circumstellar disk, 150, 153, 154, 158, 160, 163, 167–170, 173, 175–182
CNO tri-cycle, 80, 88, 89
COBE, 11, 12, 16, 17, 19, 23, 43
cold dark matter, 20–22
comets, 177, 192, 195, 197, 199
common ancestor, 190, 192, 193
convection, 72, 121, 164
core, 33, 44, 74, 79, 81–84, 99–101, 114, 115, 120–122, 130, 162
cosmic microwave background, 10–13, 15, 18, 43
cosmic rays, 68, 72, 88, 89, 201, 205

cosmological constant, 7, 8, 15, 16, 21, 124, 125
cosmological principle, 2
cosmology, 6, 8, 22, 76, 102
Crab Nebula, 130, 131
critical density, 6–11, 13, 14, 22
Cygnus X-1, 139

D

dark energy, 8, 15, 21, 22, 124–126
dark matter, 10, 11, 13, 18, 20–22, 45–47, 72, 76, 91, 95, 125
degenerate, 81, 82, 84, 89, 90, 101, 113
density fluctuations, 11, 19
deuterium, 12, 13, 67, 76, 77, 80, 88, 164
dipole, 10, 11, 129
disk accretion, 160, 161, 167–169, 173, 176, 182
DNA, 81, 104, 192, 193, 197
Doppler shift, 4, 5, 28, 111, 155
dust, 27, 30, 32, 33, 38, 39, 72, 74, 91, 92, 99, 100, 106, 116, 152–154, 158, 159, 160, 161, 171, 178, 181

E

Einstein, Albert, 7, 15, 21, 40, 45, 75, 99, 124, 125, 128, 134, 135
electromagnetic, 4, 16, 45, 80, 103, 110, 115, 120, 122, 133, 142
electrons, 2, 6, 12, 20, 39–41, 45, 66, 67, 69, 70, 76, 81, 86, 100, 101, 104, 111, 114, 118, 121, 129
Enceladus, 195, 202, 203
entropy, 149, 182
escape velocity, 7, 135
eukarya, 192, 193
Europa, 202
Evans, Reverend Robert, 107
event horizon, 136–138, 140
event horizon telescope (EHT), 140
exoplanets, 149, 178, 189, 203–205

expansion of the Universe, 1, 2, 7, 15, 21, 28, 45, 46, 56, 57, 77, 124

F

fossils, 194, 195, 197
Fowler, William, 71, 74
free-fall, 156–158, 166
Friedmann, A. A., 75

G

galaxies, 1–8, 10–13, 18–20, 22, 23, 25–61, 63, 66, 70–73, 88, 90–95, 102–104, 106–108, 112, 113, 116, 118, 122, 123, 125, 127, 133, 137, 139, 140–142, 144, 150, 165, 166, 170, 172, 189, 190, 200, 206
Galilei, Galileo, 26
gamma-ray, 12, 87, 95, 112, 117–119, 121, 122, 133, 138, 142
Gamow, George, 13, 75, 76, 86, 125
Garcia, Francisco, 106
G-dwarf, 72, 93, 94
Ghez, Andrea, 39, 42
Gladders, Mike, 53
grand unified theories, 16, 21, 119
gravitational collapse, 11, 95, 134, 135, 150, 155
gravitational instabilities, 150, 168, 169, 172, 173, 176, 179–182
gravitational waves, 87, 122, 128, 134, 142–144
Great Attractor, 10
GW 150914, 142
GW 170817, 142

H

habitable zone, 177, 178, 204
Haldane, J. B. S., 195
half-life, 12, 118, 119
Hawking, Stephen, 138
heavy bombardment, 193, 194, 202

Index

helium, 13, 14, 32, 33, 63, 64, 66, 68, 70, 72, 74–82, 84, 85, 88, 90, 95, 99, 100, 101, 103, 104, 111, 113–115, 122, 141, 177
helium-burning, 80–83, 87, 88
Herman, Robert, 76
heterotrophy, 196
Hewish, Antony, 127, 134
HL Tau, 162, 163
horizon, 14, 18, 22, 55
hot dark matter, 20, 21
Hoyle, Fred, 71, 74
Hubble constant, 1, 6–8, 15, 19, 22, 77
Hubble, Edwin, 8, 26–29, 31, 34, 125
Hubble Space Telescope (HST), 36, 37, 55, 58, 59, 73, 108, 126, 132, 140
Hulse, Russell, 134
hydrogen burning, 68, 80, 81, 87–89, 100, 162
hydrothermal vents, 198

I

IMF (Initial Mass Function), 71, 150, 165, 182
inflation, 15, 16, 17, 21–23
interstellar medium, 82, 89, 100, 102, 106, 130, 165, 166, 178, 195, 207
iron, 32, 34, 60, 64, 70, 79, 80, 83–85, 86–88, 91, 94, 104, 112, 114, 118, 119, 140, 197
isothermal, 10, 15, 159
isotopes, 13, 14, 66–68, 77, 84–86, 88–90, 118, 194, 195

J

Jupiter, 95, 175, 177, 180, 199, 202

K

Kant, Immanuel, 26
Kepler, Johannes, 116
kilonova, 104, 123, 144
Kormendy, John, 42

L

Laplace, Pierre Simon, 135
Large Magellanic Cloud, 102, 103, 116, 117
lichens, 198
LIGO, 87, 142–144
lithium, 14, 76–78, 88, 89, 103
Lundmark, K., 84
Lynden-Bell, Donald, 40, 42

M

M51, 107
M81, 106
M83, 106
M87, 41, 42, 140, 141
Madau diagram, 50, 51, 53
magnesium, 32, 34, 60, 64, 79, 89, 102, 112
magnetic field, 92, 129, 130, 132, 133, 150, 152, 154–157, 163, 164, 166, 167, 174, 175
main sequence, 52, 81, 89, 99, 113, 162, 177, 182
Mars, 114, 176, 193, 195, 199–202, 205, 206
Mercury, 95, 100, 176, 179, 193
metallicity, 206, 207
meteorites, 75, 85, 89, 195, 197, 200
methane, 196, 203
microbial, 194, 197, 198, 200, 204–206
Milky Way, 2, 10, 17, 26–35, 39, 40, 42, 43, 46, 48, 52, 56, 60, 72, 90, 91, 94, 95, 106, 113, 116, 118, 122, 127, 133, 140, 150
Miller, Stanley, 195, 196
Minkowski, Rudolph, 111
Mitchell, John, 135
molecular cloud, 33, 150, 153–159, 166, 181, 182
Moon, 14, 57, 74, 140, 193, 205
Mount Wilson, 26, 34, 111

229

N
Neptune, 167, 177, 179, 180, 199
neutralino, 21
neutrinos, 12, 13, 20, 21, 70, 76, 80, 83, 114, 119–121, 126, 131, 142, 144
neutrons, 12, 13, 45, 66–68, 70, 76–78, 80, 84–88, 99, 100, 114, 120, 122, 126, 134, 144
neutron star, 49, 102, 104, 114, 119–121, 123, 126, 128–130, 132–136, 139, 142, 144, 145, 182
NGC 4258, 140, 141
NGC 4526, 107, 108
nitrogen, 32, 34, 60, 64, 68, 80–82, 88, 89, 100, 202, 203
nova, 72, 89, 90, 94, 102, 104, 113, 123, 139
nuclear fusion, 33, 40, 44, 75, 99, 102, 112, 114, 141, 153, 159, 181
nuclear reactions, 64, 66, 67, 69, 70, 72, 77, 78, 81, 82, 85, 88, 89, 92, 100, 103, 112, 114, 115
nucleosynthesis, 66, 71, 72, 75–78, 85, 88, 89, 94, 112, 115, 118, 119

O
obliquity, 205
Oemler, Gus, 53
opacity, 72, 118, 119
Oparin, A. I., 195
Oppenheimer, J. R., 126
organic, 64, 68, 195–198, 200–203
Orion Nebula, 27, 104, 105
oxygen, 32, 34, 60, 64, 66, 70, 78, 79, 80–83, 85, 88, 89, 91, 100–102, 104, 112, 114, 193, 197, 198, 204, 206, 207

P
Palomar Observatory, 40, 108, 111
panspermia, 196

Paranal Observatory, 116, 151
Payne, Cecilia, 79, 91, 93
Perlmutter, Saul, 125
phase transition, 16
photosynthesis, 194, 197, 198, 200
phylogenetic, 192, 193
planetary nebulae, 82, 83, 88, 100, 101, 104
planetesimals, 176, 179, 180, 193
planet formation, 149, 150, 158, 173, 176, 177–180, 206
Pluto, 177
Poggianti, Bianca, 53
Population III star, 91
positrons, 12, 120, 129
pre-main-sequence, 153, 164, 167
primordial soup, 196
proteins, 64, 81, 190, 192, 193, 197
proton-proton (or p-p) chain, 80
protons, 12, 13, 45, 66–68, 70, 76–78, 80, 84–89, 99, 114, 119, 120, 142, 165
protostars, 33, 153, 158, 159, 161, 162, 170
protostellar collapse, 157, 158
protostellar phase, 153, 167
Puckett, Tim, 107
pulsar, 127–134

Q
quasars, 4, 40–42, 59, 94, 141

R
red giant, 52, 73, 81, 87, 88, 100, 101, 113, 132
redshift, 4–7, 46, 55, 56, 122, 123
reducing, 195, 197
rest mass, 121, 141
Riess, Adam, 125
RNA, 192, 193, 197
Ryle, Martin, 134

Index

S
Sargent, Wallace, 41
Saturn, 170, 177, 199, 202, 203
Schmidt, Brian, 125
Schmidt, Maarten, 40
Schwarzschild, Karl, 136
Schwarzschild radius, 136
sedimentary deposits, 194
SETI, 204, 206
Shelton, Ian, 116
shock waves, 102, 114, 160
silicon, 32–34, 60, 79, 80, 83, 85, 104, 111, 112, 208
singularity, 136, 137
Small Magellanic Cloud, 116
SN 1987A, 102, 116–121, 126, 131, 132, 142, 144
solar system, 2, 10, 66, 69, 85, 89, 90, 95, 99, 104, 122, 123, 134, 167–170, 173, 175–177, 179, 189, 190, 194, 195, 198, 199, 200, 203, 205
spallation, 89
spectral energy distributions, 161, 162, 167, 168, 169, 181
star clusters, 30, 34, 35, 49, 72, 92, 142
star formation, 28, 31–35, 44, 47–55, 57, 59–61, 71, 72, 92, 93, 112, 113, 149–153, 156, 157, 159, 166, 167, 170, 176, 181, 182, 206
steady-state model, 8, 10, 15, 69
stellar populations, 34, 35, 165
stellar winds, 100, 104, 114, 153, 162
strong nuclear force, 16
super Earths, 177
supergiant, 81, 83, 87, 114, 117, 118
supernova, 21, 34, 49, 55, 60, 72–74, 80, 83–85, 89, 87, 91, 93, 102–108, 110–119, 121–126, 130, 132, 134, 135, 144, 145, 205
supernova remnant, 74, 104, 105, 130, 131

T
Tarantula nebula, 103, 116
Taylor, Joseph H., 127, 134
technetium, 88
technological civilizations, 204, 205
terrestrial planets, 177, 179
thorium, 64, 67, 87
Thorne, Kip, 142
tidal forces, 137
time dilation, 137
Tinsley, Beatrice Muriel Hill, 70, 71, 91
Titan, 195, 203
transits, 203–205
T Tauri phase, 153, 168
T Tauri star, 154, 167
turbulence, 35, 72, 140, 149, 150, 152, 154–157, 166, 167, 173–176, 179

U
uranium, 64, 67, 85, 87, 104
Uranus, 177, 199

V
V404 Cygni, 139
vacuum energy density, 16, 21
Vela nebula, 130
Venus, 176, 199, 205
Viking landers, 200, 201
VIRGO, 142, 144
Virgo cluster, 41, 107, 140
Volkoff, G., 126
Vulcani, Benedetta, 53

W
water, 33, 59, 63, 64, 67, 95, 118–120, 137, 177, 178, 197–205, 207, 208
weakly interacting massive particle (WIMP), 20
weak nuclear interactions, 12, 16
Weiss, Rainer, 142

white dwarf, 49, 82–85, 88–90, 101, 102, 112–115, 121, 126, 128, 139, 182
WISE, 30, 31
wormhole, 145

X
X-ray binaries, 139, 142
X-ray nova, 139

Y
Young, Peter, 41

Z
Zwicky, Fritz, 112, 115, 126

Lightning Source UK Ltd.
Milton Keynes UK
UKHW022051170820
368401UK00005B/85